Contrast Sensitivity of the

HUMAN EYE

and Its Effects on Image Quality

Contrast Sensitivity of the

HUMAN EYE

and Its Effects on Image Quality

Peter G. J. Barten

SPIE Optical Engineering Press

A Publication of SPIE—The International Society for Optical Engineering
Bellingham, Washington USA

Library of Congress Cataloging-in-Publication Data

Barten, Peter G. J.
    Contrast sensitivity of the human eye and its effects on image quality / Peter G. J. Barten.
        p. cm.
    Originally published: Knegsel : HV Press, 1999.
    Thesis (doctoral) — Technische Universiteit Eindhoven, 1999.
    Includes bibliographical references and index.
    ISBN 0-8194-3496-5 (hardcover)
        1. Contrast sensitivity (Vision). 2. Contrast sensitivity (Vision) — Mathematical Models.
    I. Title.
QP481.B245   1999
612.8'4—dc21   99-40877
                                                                                    CIP

Published by

SPIE—The International Society for Optical Engineering
P.O. Box 10
Bellingham, Washington 98227-0010
Phone: 360/676-3290
Fax: 360/647-1445
Email: spie@spie.org
WWW: http://www.spie.org/

Printed in the United States of America.

to my wife Tineke,
and my children Koen, Yvonne and Marianne

Contents

Preface

In this book, a new model is given for the contrast sensitivity of the human eye with which a large number of published measurements can be explained. Furthermore, a metric is given for the calculation of the perceived quality of an image from the physical parameters of the image and the psychophysical parameters of the human visual system. The book represents the comprehensive results of about ten years of investigation in these areas. The contrast sensitivity model is based on the assumption that it is determined by the presence of internal noise in the visual system. First a fundamental mathematical analysis is given for the general properties of image noise and for the effects of noise on the perception threshold of the visual system. The effect of internal noise on contrast sensitivity is further elaborated in following chapters for various aspects of the visual system. The results are given in the form of equations that can easily be used for practical application. They are compared with a large number of empirical data. The last chapters of the book are devoted to the effect of contrast sensitivity on perceived image quality. In this part, a model is given for the nonlinear behavior of the visual system at suprathreshold levels of modulation and a metric is given for the description of image quality with the aid of the physical parameters of the imaging system and the psychophysical parameters that can be derived from the contrast sensitivity. In the last chapter, the effect of various parameters on image quality is treated, and several examples are given where the predicted image quality is compared with measurements.

The reason for the research on the subjects treated in this book, was the need for an objective measure of perceived image quality, which I felt during my professional work on the development of CRTs for television and computer display. It was clear that besides the physical data of the image, the contrast sensitivity of the eye plays an important role in such a measure. However, for the contrast sensitivity of the eye, which depends on luminance and field size, only a few measurements were available. Furthermore, it was not clear how the contrast sensitivity of the eye had to be combined with the physical parameters of the image to obtain a good measure for image quality. Therefore, I started an intensive study on these subjects after the end of my professional career. For the effect of resolution on image quality, I found that the nonlinear behavior of the visual system could be taken into account by applying a square-root relation between modulation and perceived image quality. Later it appeared that the so obtained image quality metric could not only be applied for the effect of resolution, but also for the effect of other parameters on image quality, like luminance and image size. For the effect of noise on image quality, I assumed that it was caused by the effect of noise on contrast sensitivity. To investigate this further, I made a study of published measurement data of the effect of various types of image noise on contrast sensitivity. After an evaluation of these results, the idea arose that the remarkable dependence of contrast sensitivity on luminance and field size could maybe be explained by the presence of internal noise in the visual system. However, to obtain a complete description of the contrast sensitivity function of the eye, still a number of additional assumptions had to be made. I tested these assumptions by comparing them with a large number of published data. Furthermore, I also tried to apply the same basic principles to other aspects of contrast sensitivity. The so obtained information appeared to be very useful for a further evaluation of a good image quality metric.

After having presented a part of my investigations in papers and in short courses, the idea arose to present the results more completely in a comprehensive book. For the first edition of this book, I chose the form of a dissertation at the Technical University of Eindhoven, because an important part of the measurements that I used for my investigations were made at the Institute of Perception Research (IPO) of this university. I was very glad that Prof. Roufs of this institute, who was in charge of the work on visual perception, was willing to act as supervisor of my dissertation. I am very grateful for the many hours he spent on reading the manuscript of the dissertation in a critical way and his suggestions for improvements. I am also very grateful for the support that I received during my investigations from Prof. van Nes of the same institute and from Dr. van Meeteren of the Institute for Perception TNO in Soesterberg. I also would like to thank Prof. Hooge and Prof. Butterweck of the Department of Electrical Engineering of the Technical University of Eindhoven for their advice on the mathematical treatment of the noise in Chapter 2, and I also would like to thank Dr. Tyler of the Smith-Kettlewell Eye Research Institute in San Francisco for his useful comments on Chapter 5 about the temporal contrast sensitivity. In particular, I would like to express my special thanks to my wife for her patience during the many hours that I spent on the manuscript of this book.

The present book is the textbook edition of the dissertation. It differs from the original version by the use of a hardcover, the addition of a subject index and a list of symbols, and by a few other changes and small text corrections that were made to adapt it to this application.

Peter Barten   August, 1999

List of symbols

Latin symbols

Symbol	Description	Unit
$a(u,v,w)$	amplitude of sinusoidal luminance variation	cd/m^2
A	available surface area per retinal cell	deg^2
	MTFA value	cycles/deg
$A(u,v,w)$	complex amplitude of sinusoidal luminance variation	cd/m^2
c	constant for nonlinear behavior of modulation	-----
	velocity of light	m/sec
	velocity of traveling wave	deg/sec
C	contrast factor	-----
C_{ab}	aberration constant of eye lens	arc min/mm
d	diameter of eye pupil	mm
	center-to-center distance of retinal cells	arc min
d'	detectability index	-----
D	field diameter	deg
e	numerical value of the natural logarithm (2.771828...)	-----
e	eccentricity	deg
e_g	constant used in density distribution of ganglion cells	deg
E	retinal illuminance	Td
$f(r)$	receptive field of spatial inhibition	deg^{-2}
$F(u)$	MTF of spatial inhibition filter	-----
	integrand of one-dimensional image quality metric	cond. dep.
$F(u,v)$	integrand of two-dimensional image quality metric	cond. dep.
$F(u,\vartheta)$	integrand of polar image quality metric	cond. dep.
$F(u,v,w)$	Fourier transform of luminance pattern	deg^2 sec cd/m^2
$G(u,w)$	MTF of spatiotemporal inhibition process	-----
h	Planck's constant	Joule sec
	vertical size of television image	deg
$h(t)$	temporal impulse response function	msec^{-1}

Symbol	Description	Unit
$h_1(t)$	impulse response function for MTF given by $H_1(w)$	msec^{-1}
$h_2(t)$	impulse response function for MTF given by $H_2(w)$	msec^{-1}
$H(w)$	MTF of temporal impulse response function	-----
$H_1(w)$	MTF of temporal processing of photo-receptor signal	-----
$H_2(w)$	MTF of temporal processing of spatial inhibition signal	-----
I	ICS value	cycles/deg
j	$\sqrt{-1}$	-----
j	flux density of photons	$\text{deg}^{-2}\ \text{sec}^{-1}$
$j(u)$	image quality contribution	jnd
J	image quality measure	jnd
	SQRI value	jnd
J'	modified SQRI value	jnd
k	signal-to-noise ratio at 50% detection probability	-----
k^*	signal-to-noise ratio at det. prob. different from 50%	-----
K	Kell factor (0.7)	-----
K	normalization factor SQF metric	-----
l	relative threshold elevation	-----
L	luminance	cd/m^2
$\bar{L}$	average luminance	cd/m^2
L'	output luminance	cd/m^2
$L_{\max}$	maximum luminance	cd/m^2
ΔL	luminance difference	cd/m^2
m	modulation	-----
m_0	modulation of reference signal	-----
m_n	average modulation of noise wave components	-----
m_{rel}	relative modulation of reference signal	-----
m_t	modulation threshold	-----
m_t'	increased modulation threshold	-----
Δm	modulation difference	-----
Δm_t	threshold of modulation difference	-----
Δm_{trel}	relative threshold of modulation difference	-----
$M(u)$	MTF of imaging system	-----
$M_{lat}(u)$	MTF of lateral inhibition process	-----
$M_{opt}(u)$	optical MTF of the eye	-----
n	number of photons	-----
	number of stages of impulse response function	-----
n_1	number of stages of the function $H_1(w)$	-----
n_2	number of stages of the function $H_2(w)$	-----
$\bar{n}$	average number of photons	-----
N	number of retinal cells per unit area	deg^{-2}
N_c	number of cones per unit area	deg^{-2}

Symbol	Description	Unit
N_{c0}	number of cones per unit area in center of retina	deg^{-2}
N_g	number of ganglion cells per unit area	deg^{-2}
N_{g0}	number of ganglion cells per unit area in center of retina	deg^{-2}
N_{max}	maximum number of integration cycles	-----
N_r	number of rods per unit area	deg^{-2}
N_v	number of visual scan lines	-----
p	detection probability	%
	photon conversion factor	photons/sec/deg^2/Td
	center-to-center distance of pixels	deg
p_{2AFC}	probability of correct response in a 2AFC experiment	%
$P(\lambda)$	spectral energy distribution of light source	Joule/sec
Q	SQF value	-----
r	radial distance on retina	arc min
s	signal strength	cond. dep.
	row spacing of retinal cells	arc min
s_0	signal strength at 50% detection probability	cond. dep.
s_g	row spacing of ganglion cells	arc min
S	contrast sensitivity	-----
t	time	sec
Δt	small variation of t	sec
T	temporal size	sec
T_e	integration time of the eye	sec
T_o	presentation time	sec
u	spatial frequency	cycles/deg
	spatial frequency in x direction	cycles/deg
u_0	spatial frequency limit of lateral inhibition process	cycles/deg
u_n	spatial frequency of the noise	cycles/deg
u_N	Nyquist limit of spatial frequency	cycles/deg
u_{max}	maximum spatial frequency	cycles/deg
	maximum spatial frequency in x direction	cycles/deg
u_{min}	minimum spatial frequency	cycles/deg
	minimum spatial frequency in x direction	cycles/deg
u_{nmax}	maximum spatial frequency of noise	cycles/deg
	maximum spatial frequency of noise in x direction	cycles/deg
u_{nmin}	minimum spatial frequency of noise	cycles/deg
	minimum spatial frequency of noise in x direction	cycles/deg
Δu	small variation of u	cycles/deg
Δu_n	small variation of u_n	cycles/deg
v	spatial frequency in y direction	cycles/deg
v_{max}	maximum spatial frequency in y direction	cycles/deg
v_{min}	minimum spatial frequency in y direction	cycles/deg

Symbol	Description	Unit
ν_{nmax}	maximum spatial frequency of noise in y direction	cycles/deg
ν_{nmin}	minimum spatial frequency of noise in y direction	cycles/deg
$V(\lambda)$	spectral sensitivity function for photopic light	-----
$V'(\lambda)$	spectral sensitivity function for scotopic light	-----
w	temporal frequency	Hz
x	integration variable	cond. dep.
	spatial variable in x direction	deg
Δx	small variation of x	deg
X	spatial size in x direction	deg
X_o	object size in x direction	deg
X_{max}	maximum integration area in x direction	deg
y	spatial variable in y direction	deg
Δy	small variation of y	deg
Y	spatial size in y direction	deg
Y_o	object size in y direction	deg
Y_{max}	maximum integration area in y direction	deg
z	integration limit of normal probability integral	-----

Greek symbols

Symbol	Description	Unit
α	relative active time of displayed luminance	-----
	constant for signal threshold of Weibull function	cond. dep.
β	steepness constant of Weibull function	-----
γ	exponent of displayed luminance variation	-----
γ_0	optimum value of γ	-----
ε	energy photon	Joule
η	quantum efficiency	%
ϑ	polar angle	deg
λ	wave length of light	nm
ν	light frequency of photon	sec^{-1}
π	numerical angle (3.1416...)	-----
σ	standard deviation	cond. dep.
	standard deviation of optical line-spread function	arc min
σ_0	standard dev. opt. line-spread function at small pupil size	arc min
σ_{ret}	part of this stand. dev. caused by discrete structure retina	arc min
σ_{00}	remaining part of this standard deviation	arc min
σ_{hor}	standard deviation of blur in horizontal direction	arc min
σ_{vert}	standard deviation of blur in vertical direction	arc min

Symbol	Description	Unit
σ_{dia}	standard deviation of blur in diagonal direction	arc min
σ_m	standard deviation of the modulation	-----
σ_n	relative standard deviation of the noise	-----
σ_n	relative standard deviation of the number of photons	-----
σ_r	relative standard deviation of the luminance	-----
τ	time constant of impulse response function	msec
τ_1	time constant of the function $H_1(w)$	msec
τ_{10}	value of τ_1 at low retinal illuminance and small field size	msec
τ_2	time constant of the function $H_2(w)$	msec
τ_{20}	value of τ_2 at low retinal illuminance and small field size	msec
$\Phi(u,v,w)$	spectral density	deg^2 sec
Φ_0	spectral density of neural noise	deg^2 sec
Φ_d	spectral density of nonwhite noise	cond. dep.
Φ_n	spectral density of white noise	cond. dep.
Φ_{ph}	spectral density of photon noise	deg^2 sec
$\Psi(u_n,u)$	weighting function of masking	-----

Remark:

In equations, non-standard units of variables have to be adapted to the standard units m, sec, deg, etc., unless otherwise specified.

Contrast Sensitivity of the

HUMAN EYE

and Its Effects on Image Quality

Chapter 1

Introduction

The eye plays an important role in our life, not only for seeing objects in the surrounding world, but also for reading letters, viewing paintings, photographs, films, etc. The visual acuity of the eye is generally regarded as the most important factor for the ability of the eye for seeing objects. The acuity of the eye is usually measured by acuity tests where single black letters on a white background have to be recognized, or where the minimum visible separation is measured of black rings with a small interrupted part (Landolt rings). These tests are used for decisions about the use of certain types of eye glasses, but give no information about several other factors that also play a role in the properties of the human visual system.

Objects can generally be better distinguished from each other or from their background, if the difference in luminance or color is large. Of these two factors, luminance plays the most important role. The content of this book will, therefore, concentrate on luminance, and color will be left out of consideration. In practice, it appears that it is not the absolute difference in luminance that is important, but the relative difference. This relative difference can be expressed by the ratio between two luminance values, which is called *contrast ratio*, or by the difference between two luminance values divided by the sum of them, which is simply called *contrast*. Both are dimensionless quantities. Objects that have only a small contrast with respect to their background are difficult to observe. The eye is more sensitive for the observation of objects, if the required amount of contrast is lower. The reciprocal of the minimum contrast required for detection is called *contrast sensitivity*.

For the investigation of the contrast sensitivity of the eye, different types of test patterns can be used. Generally, sinusoidal test patterns are used, as sinusoidal test patterns have an important advantage. According to Fourier analysis, the luminance pattern of an image can be considered as the sum of a number of sinusoidal luminance variations. The application of Fourier analysis in optics was first introduced by Duffieux (1946) and was later strongly promoted by Schade (1951-1955) for the analysis of the reproduction capability of image forming systems. He used it first for television systems, where the combined effect of cameras, signal transport and image tubes on the finally reproduced image can easily be described

1

with the aid of a Fourier analysis of the different parts of the system. Later he applied it also on the human eye as the final step in the image forming process (Schade, 1956). In this respect, the work of Campbell & Robson (1968) may also be mentioned. They stimulated the application of Fourier analysis and the use of sinusoidal test patterns for the investigation of the human visual system. Although Fourier analysis can strictly be used only for linear systems and the sensitivity of the eye is not linear, Fourier analysis can still be used for the area near the detection threshold, since the response at this level may be considered to be linear.

For a sinusoidal luminance pattern, contrast is defined by the amplitude of the sinusoidal variation divided by the average luminance. This quantity is called *modulation depth*, or shortly *modulation*. The minimum modulation required for the detection of this pattern is called the *modulation threshold*. As the contrast sensitivity is usually measured with sinusoidal luminance variations, the contrast sensitivity of the eye is generally defined as the reciprocal of the modulation threshold. The modulation threshold generally depends on the wavelength of the sinusoidal luminance variation, i.e., the distance between the maxima. The reciprocal of this wavelength is called *spatial frequency*. The contrast sensitivity is usually expressed as a function of this spatial frequency.

Apart from spatial luminance variations, temporal luminance variations can also occur in an image. The contrast sensitivity of the eye for these variations can be described in the same way as for spatial luminance variations. In this case the spatial frequency has to be replaced by the *temporal frequency*. For the investigation of the effect of temporal luminance variations, the pioneering work by de Lange (1952, 1954) and by Kelly (1961) may be mentioned. Both used Fourier analysis for the evaluation of these variations.

Contrast sensitivity is sometimes measured with a periodic non-sinusoidal luminance variation. In these cases, contrast is usually defined by the difference between the maximum and minimum luminance divided by the sum of them. This type of contrast is called *Michelson contrast*. For a sinusoidal luminance variation, the Michelson contrast is equal to the modulation. For a continuous repetition of a non-sinusoidal luminance variation, the equivalent sinusoidal modulation can be found by calculating the fundamental wave of this pattern with the aid of a Fourier analysis. For a square wave pattern, for instance, the modulation of the fundamental wave is $4/\pi$ times the Michelson contrast. This has to be taken into account at the evaluation of contrast sensitivity obtained with this type of data.

Knowledge of the contrast sensitivity function is important for the understanding of the visual properties of the eye. Contrary to the colorimetric sensitivity curves of the eye adopted as standard by the CIE (Commission International de l'Éclairage), there exists no such standard for the contrast sensitivity function of the eye. Defining such standard would be difficult because the contrast sensitivity of a luminance pattern depends in addition to the spatial or temporal frequency also strongly on luminance and field size. A practical expression for the spatial contrast sensitivity function, where these two parameters were also taken into account, has been given earlier by the author (Barten, 1990). It is an approximation formula based on contrast sensitivity measurements by van Meeteren & Vos (1972) for a large range of luminance levels and on contrast sensitivity measurements by Carlson (1982) for a large range of field sizes. Although in this way a practical solution has been given that can be used for technical applications, the given equation offers no insight in the fundamental basis of the contrast sensitivity of the eye.

The main purpose of this book is to give models for various aspects of contrast sensitivity based on fundamental assumptions about the functioning of the human eye. From these assumptions, expressions for the contrast sensitivity will be derived that give not only a qualitative description of contrast sensitivity but also a quantitative description. The models will be given in the form of mathematical equations that can easily be used for practical applications. The so obtained models will be extensively compared with published measurements. The central idea of these models is the assumption that contrast sensitivity is determined by internal noise in the visual system. A part of these models was already published by the author in a first version in earlier publications (Barten, 1992, 1993, 1995), but they have been brought together here in a final comprehensive state. For practical reasons the use of the models is restricted to *photopic* luminance conditions. These are the luminance conditions at daylight vision. Furthermore, the effect of the directional orientation of the luminance variations will be left out of consideration. In practice, most spatial contrast sensitivity measurements are made with horizontally or vertically oriented patterns. For these directions, the contrast sensitivity appears to be equal. Although the contrast sensitivity of the eye can be slightly different for intermediate directions (See, for instance, Campbell et al., 1966), the effect of orientation will be neglected, as it is usually very small. The so obtained equations will be used in the last chapters to evaluate the effect of contrast sensitivity on image quality.

In **Chapter 2**, insight will be given into the *psychometric function* with which the modulation threshold can be determined in a well-defined way. Based on the assumption that the contrast sensitivity is caused by internal noise, a formula will be given for the calculation of the modulation threshold from the noise. Furthermore, expressions will be given for the basic properties of image noise and for the limits of the visual system at the processing of the noise. These expressions will be used in the following chapters.

In **Chapter 3**, a model will be given for the spatial contrast sensitivity of the eye based on internal noise given in the visual system. For this model, additional assumptions will be made for the optical modulation transfer by the eye and the neural process of lateral inhibition. The model given in this chapter forms the basis of the models used in the other chapters.

The model given in Chapter 3 is restricted to the normal condition of *foveal vision* where the center of the object is imaged on the center of the retina. In **Chapter 4**, this model will be extended to *extra-foveal vision*. Extra-foveal vision is important in cases where objects also have to be observed that are outside the central area on which the eye is concentrated. At extra-foveal vision, contrast sensitivity is reduced because of the non-homogeneity of the retina. The eye has its maximum efficiency in the center of the retina. The extension of the model to extra-foveal vision is made by making some assumptions about the variation of the numerical constants used in the model with increasing eccentricity.

In **Chapter 5**, the model for the spatial contrast sensitivity given in Chapter 3 will be extended to the temporal domain by introducing some additional assumptions for the temporal behavior of the eye at the transport of information. In this way a combined spatiotemporal contrast sensitivity model is obtained. With this model also flicker effects will be explained. These occur, for instance, at the display of television and computer images.

The contrast sensitivity can also be influenced by the presence of noise in a displayed image. Although the most common type of noise is *white noise*, where the noise energy is equally distributed over all frequencies, *nonwhite noise* can also sometimes influence the contrast sensitivity. In **Chapter 6** a generalization will be given of the expressions for white noise given in Chapter 2. This generalization is made by the assumption of a distribution function that describes the masking of one spatial frequency by the presence of another spatial frequency.

Besides experiments with *contrast detection*, where a distinction has to be made between the object and a uniformly illuminated background, experiments are also sometimes made with *contrast discrimination*, where a distinction has to be made between two sinusoidal signals with a small difference in modulation. In **Chapter 7**, a model will be given for contrast discrimination. The model is based on the assumption that contrast discrimination can be considered as a special form of masking by nonwhite noise. With this model, the typical dipper shaped curves of the experimental results can be explained.

The last two chapters will be devoted to the effect of contrast sensitivity on image quality. Contrast sensitivity appears to play an important role in the subjective judgment of image quality. As images largely contain modulations at suprathreshold level, not only the contrast sensitivity of the eye at threshold level is important, but also the sensitivity of the eye at higher modulation levels. Although the contrast sensitivity is defined at threshold level, it is also related to the sensitivity of the eye at higher modulation levels. In **Chapter 8**, the contrast discrimination model given in chapter 7 is used to derive a model for the nonlinear behavior of the eye at suprathreshold levels of modulation. Based on this model, a measure, or *metric*, will be given for the perceived quality of an image where use is made of the linear relation between perceived image quality and the number of just-noticeable differences. This metric is called *square-root integral* or SQRI. This metric was described in previous publications by the author (Barten, 1987, 1989, 1990) without the more fundamental background that will be given in this chapter. In this chapter also an analysis will be given of the functional suitability of various image quality metrics for the description of perceived image quality.

In **Chapter 9**, the image quality metric given in the previous chapter will be used for an analysis of the effect of various parameters, like resolution, luminance, contrast etc. on image quality. The results will be comparted with published measurements of perceived image quality.

References

Barten, P.G.J. (1987). The SQRI method: a new method for the evaluation of visible resolution on a display. *Proceedings of the SID, 28*, 253-262.

Barten, P.G.J. (1989). The square root integral (SQRI): a new metric to describe the effect of various display parameters on perceived image quality. *Human Vision, Visual Processing, and Digital Display I, Proc. SPIE,* **1077**, 73-82.

Barten, P.G.J. (1990). Evaluation of subjective image quality with the square-root integral method. *Journal of the Optical Society of America A,* 7, 2024-2031.

Barten, P.G.J. (1992). Physical model for the contrast sensitivity of the human eye. *Human Vision, Visual Processing, and Digital Display III, Proc. SPIE,* **1666**, 57-72.

Barten, P.G.J. (1993). Spatio-temporal model for the contrast sensitivity of the human eye and its temporal aspects. *Human Vision, Visual Processing, and Digital Display IV, Proc. SPIE,* **1913**, 2-14.

Barten, P.G.J. (1995). Simple model for spatial frequency masking and contrast discrimination. *Human Vision, Visual Processing, and Digital Display VI, Proc. SPIE,* **2411**, 142-158.

Campbell, F.W., Kulikowski, J.J., & Levinson, J. (1966). The effect of the orientation on the visual resolution of gratings. *Journal of Physiology,* 187, 427-436.

Campbell, F.W. & Robson, J.G. (1968). Application of Fourier analysis to the visibility of gratings. *Journal of Physiology,* 197, 551-566.

Carlson, C.R. (1982). Sine-wave threshold contrast-sensitivity function: dependence on display size. *RCA Review*, **43**, 675-683.

de Lange, H. (1952). Experiments on flicker and some calculations on an electrical analogue of the foveal systems. *Physica*, **18**, 935-950.

de Lange, H. (1954). Relationship between critical flicker-frequency and a set of low-frequency characteristics of the eye. *Journal of the Optical Society of America*, **44**, 380-389.

Duffieux, P.M. (1946). L'intégral de Fourier et ses applications à l'optique. 1st edition 1946, 2nd edition 1970, Masson et Cie, Paris.

Kelly, D.H. (1961). Visual responses to time-dependent stimuli. I. Amplitude sensitivity measurements. *Journal of the Optical Society of America*, **51**, 422-429.

Schade, O. (1951-1955). Image gradation, graininess, and sharpness in television and motion picture systems. *Journal of the SMPTE*, **56**, 137-171 (1951), **58**, 181-222 (1952), **61**, 97-164 (1953), and **64**, 593-617 (1955).

Schade, O. (1956). Optical and photoelectric analog of the eye. *Journal of the Optical Society of America*, **46**, 721-739.

van Meeteren, A. & Vos, J.J. (1972). Resolution and contrast sensitivity at low luminance levels. *Vision Research*, **12**, 825-833.

Chapter 2

Modulation threshold and noise

2.1 Introduction

The modulation of a sinusoidal luminance pattern is defined by the amplitude of the sinusoidal variation divided by the average luminance. See Fig. 2.1. According to the generally used definition, the contrast sensitivity is the reciprocal of the threshold value of the modulation for the detection of the variation. Therefore, the modulation threshold plays an important role in the contrast sensitivity. In this chapter the concept of the modulation threshold will be treated as will the effects of noise on this threshold. In practice, it appears that there is not a fixed threshold below which a luminance variation is not observed and above which the luminance variation is

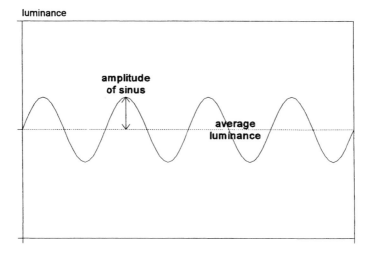

Figure 2.1: Example of a sinusoidal luminance variation. The modulation is defined by the amplitude of the sinusoidal variation divided by the average luminance. The contrast sensitivity is the reciprocal of the threshold value of the modulation for the detection of the variation.

always observed, but, instead, there is a gradual increasing probability for observing the variation. To avoid confusion, it is therefore necessary to define which modulation level is considered as threshold. A detection probability of 50% is generally used as threshold and the modulation with this detection probability is, therefore, defined as the modulation threshold. The function that describes the detection probability as a function of the signal strength is called *psychometric function*. This function is very useful to determine the modulation threshold in a well-defined way. The first part of this chapter will, therefore, be devoted to this function.

The psychometric function also gives a good understanding of the underlying detection mechanism. The statistical factors that influence the shape of the psychometric function may be considered to be caused by noise. This noise partly consists of noise generated in the visual system, which is called *internal noise*, but can also partly consist of noise that is already present in the observed image, which is called *external noise*. The basic properties of the noise will be treated in this chapter and equations will be given for the calculation of the modulation threshold from the data of the noise. The given equations will further be used in the following chapters for the calculation of the contrast sensitivity.

2.2 Psychometric function

The psychometric function gives the detection probability as a function of the signal strength. Fig. 2.2 shows an example of this function for a detection experiment by Foley and Legge (1981) with a sinusoidal luminance pattern with a spatial frequency of 2 cycles/deg. In this figure the detection probability is plotted as a function of the modulation. The continuous curve drawn through the data points is a cumulative Gaussian probability function. It has already been known for more than hundred years that the psychometric function has generally the form of this function. See, for instance, Guilford (1954, pp. 3 and 126) and Le Grand (1968, pp. 237-238). After Sir Francis Galton (1822-1911), who introduced the use of the normal probability integral for this purpose, the curve is often called Galton's ogive. The function can be described by the following expression:

$$p(s) = \frac{1}{\sigma\sqrt{2\pi}} \int_{-\infty}^{s} e^{-\frac{(x-s_0)^2}{2\sigma^2}} \, dx \qquad (2.1)$$

where p is the detection probability, s is the signal strength, s_0 is the signal strength where the detection probability is 50%, x is an integration variable, and σ is the standard deviation of the Gaussian distribution on which the psychometric function is based. For a sinusoidal luminance pattern, the signal strength s is equal to the modulation m and the signal strength s_0 is equal to the modulation threshold m_t. In

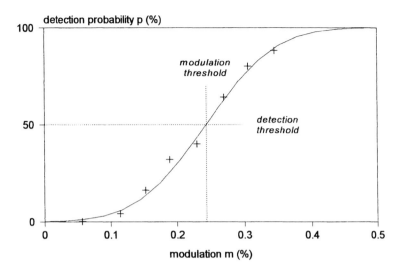

Figure 2.2: Example of a psychometric function for a sinusoidal luminance pattern with measured data from Foley & Legge (1981). A detection probability of 50% is defined as detection threshold. The modulation with this detection probability is called modulation threshold. The continuous curve through the measured data is a cumulative probability distribution function.

practice, the minimum value of s is zero. A remarkable phenomenon of the description of the psychometric function by this equation is the fact that the detection probability for a signal strength zero is not zero. However, this corresponds with the situation that a signal is observed when no signal is present, which sometimes really occurs in practice. This situation is called *false alarm*. Some investigators use the logarithm of the signal strength as independent variable for the psychometric function. Zero signal strength then corresponds with negative infinity on the log scale. This excludes the possibility of false alarm and is therefore not in agreement with practical experience.

Fig. 2.3 shows a plot of the probability density distribution. The shaded area in this figure indicates the detection probability for a signal with strength s and the double shaded area indicates the probability for false alarm. The signal strength s_0 at the maximum of the Gaussian distribution corresponds with a detection probability of 50% which is generally defined as threshold. At this probability the slope of the psychometric function reaches a maximum. The shape of this function may be explained by assuming that the detection process is subject to statistical variations that have a Gaussian distribution. See, for instance, Thurstone (1927). These variations may be considered to be caused by internal noise. The causes of this noise will be treated in more detail in the following chapter.

Eq. (2.1) can also be written in the form of the well-known *normal probability*

probability density dp/ds

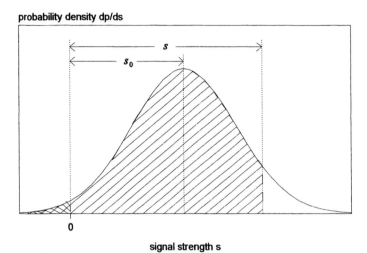

signal strength s

Figure 2.3: Probability density distribution for the detection of a signal with strength s. The shaded area indicates the detection probability $p(s)$. The double shaded area indicates the probability for false alarm $p(s=0)$.

integral

$$p(z) = \frac{1}{\sqrt{2\pi}} \int_{-\infty}^{z} e^{-\frac{x^2}{2}} \, dx \qquad (2.2)$$

where

$$z = \frac{s - s_0}{\sigma} \qquad (2.3)$$

Crozier (1935) experimentally found that the ratio between s_0 and σ is substantially constant over a wide range of signal conditions. Therefore, this property is often called *Crozier's law*. Based on this law, a constant k can be introduced by the following expression:

$$k = \frac{s_0}{\sigma} \qquad (2.4)$$

This constant may be considered as the signal-to-noise ratio required for detection. The constant k has already been used by Rose (1948) to relate the luminance threshold to the external noise present in an image. Schade (1956) found that k was in the range from 1.5 to 4.3. Roufs (1974b, p. 875) found k values ranging from 2.3 to 4 from a large number of measurements published by several investigators. (He expressed these values in fact in the inverse of k, which he called Crozier coefficient.) In practice, k can be different for different subjects and can also be different at different times of a repeated experiment. It is assumed here, that k is about 3. This

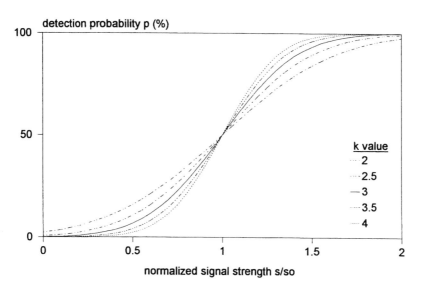

Figure 2.4: Normalized psychometric functions for different values of k.

value was also used for the calculation of the curve through the data points in Fig. 2.2. The value $k = 3$ seems quite high for a signal-to-noise ratio required for detection. However, one should consider that for a high value of k the probability of false alarm is small. For $k = 3$ the probability of false alarm is only 0.13%. During evolution, the visual system has probably developed in a direction where false alarm was avoided. This went, however, at the expense of the sensitivity for detection.

Introducing k in Eq. (2.3) gives

$$z = k\,(s/s_0 - 1) \tag{2.5}$$

From this relation follows that the psychometric function can be plotted as a normalized function of s/s_0 with k as parameter. Fig. 2.4 shows normalized plots of the psychometric function for different values of k. As can be seen, the steepness of the curves depends on the value of k. By comparing actual measurement of detection probability with these curves, the value of k can easily be determined. k can also be determined together with the threshold signal s_0 by a linear regression of s with the inverse z of the normal probability integral given by Eq. (2.2). In the past this regression was often made by plotting the detection probability p as a function of the signal strength s on probability paper. In the following, some methods will be treated that are used by investigators to determine the detection threshold. They are given here for the interpretation of the measurement results that will be used in the following chapters.

The psychometric function is sometimes approximated by a Weibull function

(Weibull, 1951). This function can be useful in the case that several processes would play a role in the detection process (Quick, 1974). The Weibull function is given by

$$p(s) = 1 - e^{-(s/\alpha)^\beta} \tag{2.6}$$

where α is the signal strength for a detection probability of $1 - 1/e = 0.632$ and β is an exponent that determines the steepness of the psychometric function. This exponent is comparable to the constant k. Sometimes α is used as signal threshold. The signal strength s_0 where the detection probability is 50% can be calculated from α if β is known:

$$s_0 = (\ln 2)^{1/\beta} \alpha \tag{2.7}$$

The slope of the Weibull function for $s = s_0$ can be calculated from Eq. (2.6) by using Eq. (2.7):

$$\left(\frac{dp}{ds} \right)_{s = s_0} = \frac{\ln 2}{2} \frac{\beta}{s_0} \tag{2.8}$$

whereas the slope of the cumulative probability integral for $s = s_0$ can be calculated from Eqs. (2.1) and (2.4), which gives

$$\left(\frac{dp}{ds} \right)_{s = s_0} = \frac{1}{\sqrt{2\pi}} \frac{k}{s_0} \tag{2.9}$$

By making the slope of the psychometric function equal to that of the Weibull function for $s = s_0$, a good fit with the Weibull function can be obtained. From Eqs. (2.8) and (2.9), it follows that in this case

$$k = \sqrt{\pi/2} \ln 2 \, \beta = 0.87 \, \beta \tag{2.10}$$

This relation can be used to calculate k from measured values of β. Fig. 2.5 shows a comparison of the psychometric function with the Weibull function calculated under this condition for three different values of k. From the figure, it can be seen that the Weibull function forms a good approximation of the psychometric function, especially for $k = 3$. It can also be seen that the Weibull function excludes the possibility of false alarm.

The psychometric function is usually measured with the *method of constant stimuli* where each point of the function is the result of a constant number of about 100 presentations of the same signal strength. Besides the value of the detection threshold, the value of k is also obtained. However, this method is very time consuming. By using some well chosen values of the signal strength around a value with a detection probability of 50%, the required effort can be reduced, but is still considerable. Therefore, in practice only the value of the detection threshold is often measured. The most simple method to obtain this threshold is the *method of adjustment*, where the signal strength is varied until the signal can just be observed. This method is quicker than other possible methods, but is less reliable, because the threshold criterion is not very well defined. An estimate of the possible error of this method can

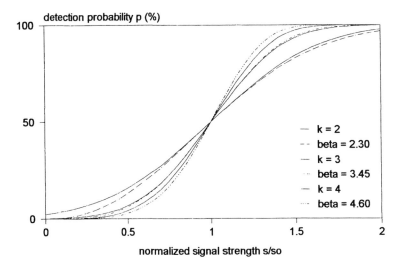

Figure 2.5: Comparison of the psychometric function for $k = 2, 3$, and 4 (solid curves) with the Weibull function for $\beta = 2.3, 3.45$, and 4.6, respectively (dashed and dotted curves).

be obtained from measurements by van Nes & Bouman (1967) that will be treated in section 3.9.4 of Chapter 3. They used the method of adjustment with a lower and a higher limit as detection criterion. From a list of the measurement data given by van Nes (1968), it appears that the lower and higher limits differed approximately 12% from the average of both. From Eqs. (2.2) through (2.4) and the k value of these measurements can be derived that the lower and higher limits in this experiment corresponded with 37% and 63% probability of detection, respectively. This gives an indication about the accuracy that can be obtained with the method of adjustment.

Another often used method is the *two-alternative forced choice* method or 2AFC method. The observer has to tell which one of two presented stimuli contains the test signal. If he cannot make a distinction, he has to make a guess. Taking into account that the guessing has a probability of 50% to be correct, the total probability of correct response in this experiment is given by

$$P_{2\mathrm{AFC}}(s) = \frac{1}{2} + \frac{1}{2}\,p(s) \tag{2.11}$$

This means that 75% correct response corresponds with a detection probability of 50%. More generally, x% correct response corresponds with a detection probability of $2(x\text{-}50)$%. This relation has been used to calculate the detection probability of the measurement data shown in Fig. 2.1, which were originally given as the results of a 2AFC experiment. To arrive quickly at the situation of 75% correct response, a *staircase procedure* is often used where the presented signal strength is changed depending on the results of the previous observations. Sometimes staircase procedures are used that give a result corresponding with 79%, 84%, or 90% correct response, instead of 75%. This corresponds with a detection probability of 58%, 68%, or 80% instead of 50%. In such a case the obtained results must be corrected. The value of k obtained from the results without correction can be described by

$$k^* = \frac{s}{\sigma} \qquad (2.12)$$

where s differs from the value s_0 used in Eq. (2.4). From Eq. (2.3) follows

$$z = \frac{s - s_0}{\sigma} = k^* - k \qquad (2.13)$$

so that the corrected value of k is given by

$$k = k^* - z \qquad (2.14)$$

The value of z in this expression can be calculated by using Eq. (2.2) in inverse form. For a correct response of 79%, 84%, and 90% in a 2AFC experiment, the obtained z value is 0.20, 0.47, and 0.84, respectively. The value of k for such an experiment has to be corrected with one of these values.

The results of a 2AFC experiment are sometimes described with the *detectability index* d'. The quantity d' is a measure for the detectability of a signal and is equal to the distance between the means of two distributions divided by the square root of the sum of squares of their standard deviations. It was originally designed for acoustical experiments. See Tanner & Birdsall (1958). For a 2AFC experiment d' is defined by

$$d' = \sqrt{2} \, p_{2AFC}^{-1}(z) \qquad (2.15)$$

where $p_{2AFC}^{-1}(z)$ is the inverse of the normal probability integral given by Eq. (2.2), but using the probability of correct response in a 2AFC experiment, instead of the detection probability. See, for instance, Legge (1984) and Pelli (1985). According to this definition the probability of correct response is given by

$$p_{2AFC}(d') = \frac{1}{\sqrt{2\pi}} \int_{-\infty}^{\frac{d'}{\sqrt{2}}} e^{-\frac{x^2}{2}} \, dx \qquad (2.16)$$

Care should be taken in the use of d' values, because d' is not linearly related with the value of z used in Eq. (2.2). The relation between d' and z can be derived from Eqs. (2.16), (2.11) and (2.2) and is shown in Fig 2.6. In practice the value $d' = 1$ is used to characterize the threshold. This value corresponds with $p_{2AFC} = 76\%$ instead of 75%.

For a sinusoidal luminance pattern, the signal strength s is given by the modulation m of the luminance pattern and the threshold signal s_0 by the modulation threshold m_t at 50% probability of detection. For a sinusoidal luminance pattern, Eq. (2.5) can be replaced by

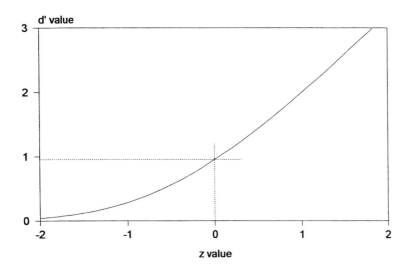

Figure 2.6: Relation between the value for d' and z for a 2AFC experiment. For $p_{2AFC} = 75\%$, $z = 0$ and $d' = 0.954$.

$$z = k\,(m/m_t - 1) \qquad (2.17)$$

and Eq. (2.4) by

$$k = \frac{m_t}{\sigma_m} \qquad (2.18)$$

where σ_m represents the probability density distribution of the modulation. The value of k appears not to be different from the value of k for other types of signals. The probability density distribution of the modulation required for detection may be assumed to be caused by noise. It is assumed here that the standard deviation σ_m of this distribution is equal to the average modulation m_n of the noise wave components, where this average modulation is defined as the RMS (= root of the mean of squares) of the noise multiplied by $\sqrt{2}$. This multiplication factor is caused by the sinusoidal definition of the modulation. According to this assumption

$$k = \frac{m_t}{m_n} \qquad (2.19)$$

or

$$m_t = k\,m_n \qquad (2.20)$$

This equation forms the basis of the contrast sensitivity model that will be given in the following chapters. It says that the modulation threshold is a factor k larger than the average modulation of the noise wave components. The noise need not consist only of internal noise, but can also partly consist of external noise. In the following section will be treated how m_n can be derived from the spectral density of the noise.

2.3 Basic properties of image noise

According to Fourier analysis, an arbitrary luminance pattern $L(x,y,t)$, where x and y are the spatial dimensions and t is the temporal dimension, can be considered to consist of sinusoidal luminance components with spatial frequency u and v, and temporal frequency w. This is also valid for the difference between the luminance $L(x,y,t)$ and the average luminance $\bar{L}$ of the pattern. In complex notation, this difference can be written as

$$L(x,y,t) - \bar{L} = \int_{-\infty}^{+\infty} \int_{-\infty}^{+\infty} \int_{-\infty}^{+\infty} F(u,v,w)\, e^{\,j2\pi(ux + vy + wt)}\, du\, dv\, dw \qquad (2.21)$$

where $j = \sqrt{-1}$ and $F(u,v,w)$ is the Fourier transform of $(L(x,y,t) - \bar{L})$ given by

$$F(u,v,w) = \int_{-\infty}^{+\infty} \int_{-\infty}^{+\infty} \int_{-\infty}^{+\infty} (L(x,y,t) - \bar{L})\, e^{\,-j2\pi(ux + vy + wt)}\, dx\, dy\, dt \qquad (2.22)$$

See, for instance, Papoulis (1968, pp. 89-90). For the situations that will be considered here, the luminance pattern has limited spatial and temporal dimensions given by X, Y and T, where $L(x,y,t) - \bar{L}$ is assumed to be zero outside this range. In this case, Eq. (2.22) can be written as

$$F(u,v,w) = \int_{-\frac{1}{2}X}^{+\frac{1}{2}X} \int_{-\frac{1}{2}Y}^{+\frac{1}{2}Y} \int_{-\frac{1}{2}T}^{+\frac{1}{2}T} (L(x,y,t) - \bar{L})\, e^{\,-j2\pi(ux + vy + wt)}\, dx\, dy\, dt \qquad (2.23)$$

The average luminance of this pattern is then given by

$$\bar{L} = \frac{1}{XYT} \int_{-\frac{1}{2}X}^{+\frac{1}{2}X} \int_{-\frac{1}{2}Y}^{+\frac{1}{2}Y} \int_{-\frac{1}{2}T}^{+\frac{1}{2}T} L(x,y,t)\, dx\, dy\, dt \qquad (2.24)$$

The complex amplitude $A(u,v,w)$ of the various frequency components is given by

$$A(u,v,w) = \frac{F(u,v,w)}{XYT} \qquad (2.25)$$

From the last equations, it follows that for zero spatial and temporal frequencies,

$$A(0,0,0) = 0 \qquad (2.26)$$

For the real amplitude $a(u,v,w)$ of the sinusoidal luminance components, it can be derived that

$$a(u,v,w) = |A(u,v,w)| + |A(-u,-v,-w)| \qquad (2.27)$$

From Eqs. (2.23) and (2.25), it follows that

$$|A(-u,-v,-w)| = |A(u,v,w)| \qquad (2.28)$$

so that

$$a(u,v,w) = 2\,|A(u,v,w)| \qquad (2.29)$$

For the applications considered here, instead of the amplitude of the luminance components, the modulation $m(u,v,w)$ is used, which is equal to the amplitude divided by the average luminance. For this modulation holds then,

$$m(u,v,w) = \frac{2|A(u,v,w)|}{\overline{L}} \tag{2.30}$$

An important quantity of the luminance pattern is the *relative variance of the luminance*, defined by

$$\sigma_r^2 = \frac{1}{XYT} \int_{-\frac{1}{2}X}^{+\frac{1}{2}X} \int_{-\frac{1}{2}Y}^{+\frac{1}{2}Y} \int_{-\frac{1}{2}T}^{+\frac{1}{2}T} \frac{\{L(x,y,z) - \overline{L}\}^2}{\overline{L}^2} \, dx\,dy\,dt \tag{2.31}$$

σ_r is the *relative standard deviation of the luminance* and is a dimensionless quantity. Applying Parseval's theorem on Eqs. (2.21), and (2.23), gives

$$\int_{-\frac{1}{2}X}^{+\frac{1}{2}X} \int_{-\frac{1}{2}Y}^{+\frac{1}{2}Y} \int_{-\frac{1}{2}T}^{+\frac{1}{2}T} \{L(x,y,z) - \overline{L}\}^2 \, dx\,dy\,dt = \int_{-\infty}^{+\infty} \int_{-\infty}^{+\infty} \int_{-\infty}^{+\infty} |F(u,v,w)|^2 \, du\,dv\,dw \tag{2.32}$$

See, for instance, Papoulis (1968, p. 92). This means that Eq. (2.31) can be written as

$$\sigma_r^2 = \frac{1}{XYT} \int_{-\infty}^{+\infty} \int_{-\infty}^{+\infty} \int_{-\infty}^{+\infty} \frac{|F(u,v,w)|^2}{\overline{L}^2} \, du\,dv\,dw \tag{2.33}$$

This expression shows that σ_r^2 consists of the contributions of the various frequency components of the luminance pattern. The equation can be simplified by introducing the *relative power spectral density* $\Phi(u,v,w)$ defined by

$$\Phi(u,v,w) = \frac{1}{XYT} \frac{|F(u,v,w)|^2}{\overline{L}^2} \tag{2.34}$$

For practical reasons this function will simply be called here spectral density, so the additions "relative" and "power" will further be omitted. By using this function, Eq. (2.33) becomes

$$\sigma_r^2 = \int_{-\infty}^{+\infty} \int_{-\infty}^{+\infty} \int_{-\infty}^{+\infty} \Phi(u,v,w) \, du \, dv \, dw \tag{2.35}$$

The spectral density used here is the two-sided defined spectral density (See Legge et al. 1987) contrary to the sometimes used one-sided defined spectral density. From Eq. (2.28) follows that for a simultaneous sign change of the three frequencies

$$\Phi(-u,-v,-w) = \Phi(u,v,w) \tag{2.36}$$

By introducing Eq. (2.34) in Eq. (2.30), one obtains with the aid of Eq. (2.25)

$$m(u,v,w) = 2 \sqrt{\frac{\Phi(u,v,w)}{XYT}} \tag{2.37}$$

For noise, the same expressions are valid as for a normal image. For noise, Eq. (2.35) can be written in the form

$$\sigma_n^2 = \int_{-\infty}^{+\infty} \int_{-\infty}^{+\infty} \int_{-\infty}^{+\infty} \Phi_n(u,v,w) \, du \, dv \, dw \tag{2.38}$$

where σ_n is the relative standard deviation of the noise and $\Phi_n(u,v,w)$ is the spectral density of the noise. Since for noise, the spectral density is also the same for non-simultaneous sign changes of the three frequencies, this expression can be simplified to

$$\sigma_n^2 = \int_0^\infty \int_0^\infty \int_0^\infty \Phi_n(u,v,w) \, 2du \, 2dv \, 2dw \tag{2.39}$$

where the integration is extended only over positive frequencies.

The commonly occurring type of noise is white noise. White noise is characterized by the property that the (power) spectral density is constant over all frequencies. However, white noise is strictly only a theoretical quantity. In practical situations, white noise occurs as an approximation, valid within a given limited frequency range. Outside this range the spectral density decreases to zero. For white noise with an idealized rectangular spectrum, Eq. (2.39) becomes

$$\sigma_n^2 = \int_{u_{nmin}}^{u_{nmax}} \int_{v_{nmin}}^{v_{nmax}} \int_{w_{nmin}}^{w_{nmax}} \Phi_n \, 2du \, 2dv \, 2dw \tag{2.40}$$

where u_{nmin}, u_{nmax}, etc. are the minimum and maximum spatial and temporal frequencies, which are defined here as positive values. The minimum frequencies are usually zero. In this case

$$\Phi_n = \frac{\sigma_n^2}{2u_{nmax} \, 2v_{nmax} \, 2w_{nmax}} \tag{2.41}$$

whereas for nonzero minimum frequencies, u_{nmax} has to be replaced by $(u_{nmax} - u_{nmin})$, etc.. With this equation, the spectral density of the noise can be calculated when the relative standard deviation of the noise is given. If the noise is measured with samples with dimensions Δx, Δy and Δt and truncation errors are neglected, $u_{nmax} = 1/(2\Delta x)$, $v_{nmax} = 1/(2\Delta y)$, and $w_{nmax} = 1/(2\Delta t)$, as a minimum of two samples is needed to obtain one cycle. From Eq. (2.41) follows that in this case

$$\Phi_n = \sigma_n^2 \, \Delta x \, \Delta y \, \Delta t \tag{2.42}$$

For the average modulation m_n of the noise wave components, one obtains with the aid of Eq. (2.37)

$$m_n = 2 \sqrt{\frac{\Phi_n}{XYT}} \tag{2.43}$$

With the aid of this equation, m_n can be calculated from the spectral density of the noise. In this equation, X, Y, and T are the spatial and temporal size of the object covered with noise. From this relation follows that for the idealized situation of white noise, the value of m_n is equal for all frequencies.

Noise can be internal noise, present in the visual system, or external noise present in the observed image, or a combination of both. External noise can be purely static, like, for instance, grain noise in a photographic picture. Then no temporal noise is present and the factor T in Eq. (2.43) and w_{nmax} in Eq. (2.41) have to be omitted. Sometimes the spatial noise is only present in one dimension. Then the factors Y and v_{nmax} in these expressions have to be omitted. In these situations the spectral density has a different dimension. However, by expressing the effects of the noise in the dimensionless quantity m_n, these situations can easily be compared with each other.

2.4 Effect of noise on modulation threshold

The effect of external noise on the modulation threshold can also reveal much about the behavior of the visual system with respect to internal noise. External noise can easily be measured, whereas these possibilities do not exist for internal noise.

Van Meeteren & Valeton (1988) empirically found that the effect of external noise on the modulation threshold can be described by the following equation:

$$m_t' = \sqrt{m_t^2 + c^2 m_n^2} \qquad (2.44)$$

where m_t' is the modulation threshold with external noise, m_t is the modulation threshold without external noise, m_n is the average modulation of the external noise, and c is a dimensionless constant. From Eq. (2.20) one would expect that the constant c is equal to k. To verify this, we analyzed investigations of contrast sensitivity measurements with and without noise published by various authors. These measurements were made with different types of noise: one-dimensional static noise, one-dimensional dynamic noise, two-dimensional static noise, and two-dimensional dynamic noise. See Table 2.1. For these measurements m_n has been calculated with Eqs. (2.43) and (2.41) and c with Eq. (2.44). The results are given in the table. In some cases the results differ considerably from the expected value 3 mentioned in section 2.2. However, during a previous study (Barten, 1991) we found that a value closer to 3 can be obtained by the introduction of some limits to the values of X, Y and T used in Eq. (2.43) which represent the integration area given by Eq. (2.23). These limits are caused by the limited capability of the eye to perform this integration.

Table 2.1 Values of c and k calculated from measurements with external noise.

author	noise type	lumin-ance (cd/m²)	object size (deg²)	viewing time (sec)	spatial freq. (c/deg)	c	k
Strohmeyer et al. (1972)	1-d dyn.	16	2.5×1	no limit	1.77 5 10	3.0 3.2 4.0	2.9 2.8 2.9
Pelli (1985)	2-d dyn.	300	4×4	0.07	4	4.3[*]	2.9[*]
Thomas (1985)	1-d stat.	65	3×3	0.5	6.25 8.75	3.0 4.1	2.5 3.0
Legge et al. (1987)	2-d dyn.	340	1×1	0.16	2	5.8	4.5
van Meeteren et al. (1988)	2-d stat.	100	1×1	0.2	1 2 4 9 18	3.6 3.1 3.1 4.1 4.0	3.6 3.0 3.0 3.5 2.5

[*] After correction of the spectral density with a factor 2^3 due to the use of a different definition of the spectral noise density.

For T, it has been assumed that with a presentation time T_o of the object and an integration time T_e of the eye, the shortest of both has to be used. This may be expressed by the following expression:

$$T = \left(\frac{1}{T_o^2} + \frac{1}{T_e^2} \right)^{-0.5} \tag{2.45}$$

This expression also holds when T_o and T_e are about equal. For the integration time of the eye, Schade (1956) mentioned a value of 0.1 sec for nearly all luminance levels, and only a slightly higher value at very low luminance levels. Although values varying from 15 msec to 300 msec can be found in published papers for different types of conditions (Barlow, 1958; Roufs, 1974a), a constant value of 0.1 sec will be used here under all conditions. This value may be considered as a practical average that appears to give a best fit with the data for nearly all conditions.

In a similar way, it may be assumed that the spatial dimensions X and Y are limited by a maximum angular size. This limitation may be expressed by the following equations:

$$X = \left(\frac{1}{X_o^2} + \frac{1}{X_{max}^2} \right)^{-0.5} \tag{2.46}$$

and

$$Y = \left(\frac{1}{Y_o^2} + \frac{1}{Y_{max}^2} \right)^{-0.5} \tag{2.47}$$

where X and Y are expressed in angular size for the eye, X_o and Y_o are the angular size of the object in the x and y direction, respectively, and X_{max} and Y_{max} are the maximum angular dimensions of the integration area. It may further be assumed that $X_{max} = Y_{max}$. From measurements by Carlson (1982), which will be treated in section 3.9.11 of Chapter 3, it can be derived that X_{max} is about 12°.

However, from several published measurements (Hoekstra et al., 1974, Savoy & McCann, 1975, Howell & Hess, 1978, Virsu & Rovamo, 1979, Robson & Graham, 1981, Jamar & Koenderink, 1983), it appears that there is also a limit of the integration area formed by a maximum number of cycles. If N_{max} is the maximum number of cycles, the maximum angular size caused by this limit is N_{max}/u where u is the spatial frequency. The combined effect of this limit with the last two limits may be expressed by

$$X = \left(\frac{1}{X_o^2} + \frac{1}{X_{max}^2} + \frac{u^2}{N_{max}^2} \right)^{-0.5} \tag{2.48}$$

and

$$Y = \left(\frac{1}{Y_o^2} + \frac{1}{Y_{max}^2} + \frac{u^2}{N_{max}^2} \right)^{-0.5} \tag{2.49}$$

These expressions replace Eqs. (2.46) and (2.47). The spatial frequency u used in these expressions is the total spatial frequency, independent of orientation. As the limit formed by the maximum number of cycles is inversely proportional to spatial frequency, it mainly effects the modulation threshold at high spatial frequencies. In the mentioned papers, a maximum number of cycles ranging from 5 to 25 can be found. The large spread of this number is probably caused by the difference in measurement conditions. See, for instance, Estevez & Cavonius (1975), McCann et al. (1978), McCann & Hall (1980), van der Wildt & Waarts (1983). A number of 15 cycles will be used here for N_{max}, which appears to give a best fit with most of the published measurements.

The limitation of the integration area of the eye by a maximum number of cycles looks somewhat strange. However, this limitation is probably caused by the decrease of the contrast sensitivity with increasing distance from the center of the retina. This subject will be treated in more detail in section 4.4.2 of Chapter 4.

By using the above given expressions for X, Y, and T, corrected values of c are found that are given in the last column of Table 2.1. They may be considered to represent the actual value of k and are, therefore, indicated with k. They vary with some spread around 3, but show a systematic dependence neither on noise type, nor on object size, luminance or spatial frequency. This means that for external noise, Eq. (2.44) can be replaced by

$$m_t' = \sqrt{m_t^2 + k^2 m_n^2} \tag{2.50}$$

if the limitations for X, Y, and T given by Eqs. (2.48), (2.49) and (2.45), respectively, are used for the calculation of m_n.

For the often occurring situation that the object dimensions in x and y directions are equal, Eqs. (2.48) and (2.49) reduce to

$$X = Y = \left(\frac{1}{X_o^2} + \frac{1}{X_{max}^2} + \frac{u^2}{N_{max}^2} \right)^{-0.5} \tag{2.51}$$

If X and Y are not too much different, or if the object is circular, the same expression can be used. Then, X_o^2 has to be taken equal to the total angular area of the object.

After the author mentioned the here given equations for the spatial and temporal limits of the integration in earlier publications on this subject (Barten, 1991, 1992), a similar expression was proposed by Rovamo et al. (1993). However, they applied the spatial limitation only to the total area, instead of separately to the two dimensions of it. This makes no difference when the angular dimensions in x and y direction are nearly equal. However, when the dimensions in these directions are substantially different, Eqs. (2.48) and (2.49) will give a better fit with the measurements.

2.5 Summary and conclusions

The contrast sensitivity of the eye is defined by the modulation threshold for the detection of sinusoidal signals. In this chapter, a treatment has been given of the psychometric function by which this modulation threshold can be determined in a well-defined way and some methods has been described that are generally used for the measurement of the modulation threshold.

It is assumed here that the modulation threshold is caused by noise. The noise consists of internal noise present in the visual system, but can partly also consist of external noise present in the observed image. According to the model given here, the modulation threshold is a fixed factor k larger than the average modulation of the noise wave components. The factor k is about 3.

From the basic properties of image noise, expressions have been derived for the calculation of the average modulation of the noise wave components from the data of the noise. Furthermore, equations have been given for the maximum spatial and temporal dimensions over which the eye can integrate the information of the luminance pattern. All these equations will be used in the following chapters for a further evaluation of the contrast sensitivity.

References

Barlow, H.B. (1958). Temporal and spatial summation in human vision at different background intensities. *Journal of Physiology*, 141, 337-350.

Barten, P.G.J. (1991). Evaluation of the effect of noise on subjective image quality. *Human Vision, Visual Processing, and Digital Display II, Proc. SPIE*, 1453, 2-15.

Barten, P.G.J. (1992). Physical model for the contrast sensitivity of the human eye. *Human Vision, Visual Processing, and Digital Display III, Proc. SPIE*, 1666, 57-72.

Carlson, C.R. (1982). Sine-wave threshold contrast-sensitivity function dependence on display size. *RCA Review*, 43, 675-683.

Crozier, W.J. (1935). On the variability of critical illumination for flicker fusion and intensity discrimination. *Journal of General Physiology*, 19, 503-522.

Estevez, O. & Cavonius, C.R. (1975). Low-frequency attenuation in the detection of gratings; sorting out the artefacts. *Vision Research*, 16, 497-500.

Foley, J.M. & Legge, G.E. (1981). Contrast detection and near-threshold discrimination in human vision. *Vision Research*, 21, 1041-1053.

Guilford, J.P. (1954). Psychometric methods. 2nd edition, McGraw-Hill, New York.

Hoekstra, J, van der Goot, D.P.J, van den Brink, G, and Bilsen, F.A. (1974). The influence of the number of cycles upon the visual contrast threshold for spatial sine wave patterns. *Vision Research*, 14, 365-368.

Howell, E.R. & Hess, R.F. (1978). The functional area for summation to threshold for sinusoidal gratings. *Vision Research*, 18, 369-374.

Jamar, J.H.T. & Koenderink, J. (1983). Sine-wave gratings scale invariance and spatial integration at suprathreshold contrast. *Vision Research*, 23, 805-810.

Legge, G.E. (1984). Binocular contrast summation-I. Detection and discrimination. *Vision Research*, 24, 373-383.

Legge, G.E., Kersten, D., and Burgess, A.E. (1987). Contrast discrimination in noise. *Journal of the Optical Society of America A*, 4, 391-404.

le Grand, Y. (1968). Light, colour and vision. 2nd edition, Chapman and Hall,

London.

McCann, J.J. & Hall, J.A. Jr (1980). Effects of average-luminance surrounds on the visibility of sine-wave gratings. *Journal of the Optical Society of America*, **70**, 212-219.

McCann, J.J., Savoy R.L., and Hall, J.A. Jr (1978). Visibility of low-frequency sine-wave targets dependence on number of cycles and surround parameters. *Vision Research*, **18**, 891-894.

Nachmias, J. & Sansbury, R.V. (1974). Grating contrast discrimination may be better than detection. *Vision Research*, **14**, 1039-1041.

Papoulis, A. (1968). Systems and transforms with applications in optics. McGraw-Hill, New York-St. Louis-San Francisco-Toronto-London-Sydney.

Pelli, D.G. (1985). Uncertainty explains many aspects of visual contrast detection and discrimination. *Journal of the Optical Society of America A*, **2**, 1508-1532.

Quick, R.F. (1974). A vector magnitude model of contrast detection. *Kybernetik*, **16**, 65-67.

Robson, J.G. & Graham, N. (1981). Probability summation and regional variation in contrast sensitivity across the visual field. *Vision Research*, **21**, 409-418.

Rose, A (1948). The sensitivity performance of the human eye on an absolute scale. *Journal of the Optical Society of America*, **38**, 196-208.

Roufs, J.A.J. (1974a). Dynamic properties of vision-IV. Thresholds of decremental flashes, incremental flashes and doublets in relation to flicker fusion. *Vision Research*, **14**, 831-851.

Roufs, J.A.J. (1974b). Dynamic properties of vision-VI. Stochastic threshold fluctuations and their effect on flash-to-flicker sensitivity ratio. *Vision Research*, **14**, 871-888.

Rovamo, J., Luntinen, O., and Näsänen, R. (1993). Modelling the dependence of contrast sensitivity on grating area and spatial frequency. *Vision Research*, **33**, 2773-2788.

Savoy, R.L. & McCann, J.J. (1975). Visibility of low-spatial-frequency sine-wave targets dependence on the number of cycles. *Journal of the Optical Society of America*, **65**, 343-350.

Schade, O. (1956). Optical and photoelectric analog of the eye. *Journal of the Optical Society of America*, **46**, 721-739.

Stromeyer, C.F. & Julesz, B. (1972). Spatial frequency masking in vision: critical bands and spread of masking. *Journal of the Optical Society of America*, **62**, 1221-1232.

Tanner, W.P. & Birdsall, T.G. (1958). Definitions of d' and η as psychophysical measures. *Journal of the Acoustical Society of America*, **30**, 922-928.

Thomas, J.P. (1985). Effect of static-noise and grating masks on detection and identification of grating targets. *Journal of the Optical Society of America A*, **2**, 1586-1592.

Thurstone, L.L. (1927). Psychophysical analysis. *American Journal of Psychology*, **38**, 368-389.

van der Wildt, G.J. & Waarts, R.G. (1983). Contrast detection and its dependence on the presence of edges and lines in the stimulus field. *Vision Research*, **23**, 821-830.

van Meeteren, A & Valeton, J. (1988). Effects of pictorial noise interfering with visual detection. *Journal of the Optical Society of America A*, **5**, 438-444.

van Nes, F.L. (1968). Experimental studies in spatiotemporal contrast transfer by the human eye. Ph.D. Thesis, Utrecht University, Utrecht, The Netherlands.

van Nes, F.L. & Bouman, M.A. (1967). Spatial modulation transfer in the human eye. *Journal of the Optical Society of America*, **57**, 401-406.

Virsu, V. & Rovamo, J. (1979). Visual resolution, contrast sensitivity, and the cortical magnification factor. *Experimental Brain Research*, **37**, 475-494.

Weibull, W. (1951) A statistical distribution function of wide applicability. *Journal of Applied Mechanics*, **18**, 292-297.

Chapter 3

Model for the spatial contrast sensitivity
of the eye

3.1 Introduction

In the previous chapter, equations were given for the effect of noise on contrast sensitivity. In this chapter, these equations will be used for a model of the spatial contrast sensitivity of the eye. This model is based on the assumption that the contrast sensitivity is mainly determined by the internal noise generated in the visual system. For this model, additional assumptions have to be made about the optical properties of the eye and about the neural processing of the information. In this way, a quantitative description of the contrast sensitivity function will be obtained that also explains the dependence of contrast sensitivity on luminance and field size. The predictions by this model will be compared with a large number of published measurements of the contrast sensitivity. These measurements are usually made at medium and high luminance, which condition is called *photopic vision* (= daylight vision), but are sometimes also made at low luminance, which condition is called *scotopic vision* (= night vision). At photopic vision the cones act as photo-receptors, whereas at scotopic vision the rods act as photo-receptors. For practical reasons, the application of the model is restricted to photopic vision.

In the model, use will be made of the *modulation transfer function* or MTF. This function describes the filtering of the modulation by an image forming system as a function of the spatial frequency. The use of an MTF has the advantage that according to the convolution theorem, the MTFs of different parts of an image forming system can simply be multiplied with each other to obtain the total effect on the image. See, for instance, Papoulis (1968, p. 74). The MTF is based on the application of Fourier analysis and can, therefore, only be applied to linear systems. However, as the model is based on threshold signals and the system may be assumed to be linear around the threshold, nonlinearity effects may be neglected. From a comparison of the model with measured data, it appears that this neglect is justified.

3.2 Outline of the model

In the model, it is assumed that a luminance signal that enters the eye is first filtered by the optical MTF of the eye and then by the MTF of a lateral inhibition process. It is further assumed that the optical MTF is mainly determined by the eye lens and the discrete structure of the retina, and that the MTF of the lateral inhibition is determined by neural processing. For a comparison of the signal with the internal noise, Eq. (2.20) in Chapter 2 has to be modified into

$$m_t M_{opt}(u)\, M_{lat}(u) = k m_n \tag{3.1}$$

where $M_{opt}(u)$ is the optical MTF of the eye, $M_{lat}(u)$ is the MTF of the lateral inhibition process and m_n is the average modulation of the internal noise. After applying Eq. (2.43) to m_n at the right-hand side of this equation, one obtains

$$m_t M_{opt}(u)\, M_{lat}(u) = 2k \sqrt{\frac{\Phi_n}{XYT}} \tag{3.2}$$

where Φ_n is the spectral density of the internal noise and X, Y, and T are the spatial and temporal dimensions of the object, where the limited integration area of the visual system has to be taken into account by using Eqs. (2.48), (2.49), and (2.45), respectively, for these quantities.

Internal noise is partly due to photon noise caused by statistical fluctuations of the number of photons that generate an excitation of the photo-receptors, and partly due to neural noise caused by statistical fluctuations in the signal transport to the brain. Although the original image already contains photon noise before entering the eye, photon noise is not considered here as external noise, but as internal noise. This treatment might be clear from the fact that the spatial frequency components of this noise are not filtered by the lowpass filter formed by the eye lens. The spectral density of the internal noise may, therefore, be written in the form

$$\Phi_n = \Phi_{ph}\, M_{lat}^{2}(u) + \Phi_0 \tag{3.3}$$

where Φ_{ph} is the spectral density of the photon noise, and Φ_0 is the spectral density of the neural noise. In this equation, it is assumed that the photon noise is filtered together with the signal by the lateral inhibition process.

Fig. 3.1 shows a block diagram of the model. For completeness, external noise is also mentioned in this figure. External noise can, for instance, consist of display noise present in a television image, or of grain noise present in a photographic image. The spectral noise density of this external noise adds to the spectral noise density of the internal noise after multiplication by $M_{opt}^{2}(u)M_{lat}^{2}(u)$. However, in most of the cases treated in this chapter no external noise is present.

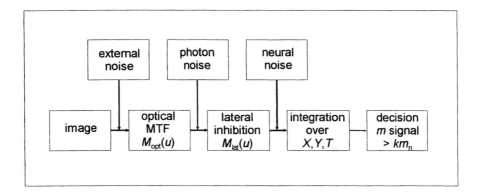

Figure 3.1: Block diagram of the processing of information and noise according to the contrast sensitivity model described here.

Insertion of Eq. (3.3) in Eq. (3.2) gives

$$m_t \, M_{opt}(u) \, M_{lat}(u) \;=\; 2\,k \; \sqrt{\frac{\Phi_{ph} \, M_{lat}^2(u) \;+\; \Phi_0}{XYT}} \qquad (3.4)$$

The contrast sensitivity S, which is the inverse of the modulation threshold m_t, is then given by

$$S(u) \;=\; \frac{1}{m_t(u)} \;=\; \frac{M_{opt}(u)}{2\,k} \; \sqrt{\frac{XYT}{\Phi_{ph} \;+\; \Phi_0/M_{lat}^2(u)}} \qquad (3.5)$$

This expression forms the basis of the given contrast sensitivity model given here. The various components of this expression will be treated in more detail in the following sections.

3.3 Optical MTF

The optical MTF used in the model does not include only the optical MTF of the eye lens, but also the effects of stray light in the ocular media, diffusion in the retina and the discrete structure of the photo-receptors. These effects have to be convolved with each other to obtain the total effect. For many convolutions in succession, the *central limit theorem* may be applied. See, for instance, Papoulis (1968, pp. 78-80). This theorem says that the total effect of several lowpass MTFs can be described by a Gaussian function. Therefore, it is assumed here that the optical MTF of the eye can be described by the following function:

$$M_{opt}(u) = e^{-2\pi^2\sigma^2 u^2} \qquad (3.6)$$

where σ is the standard deviation of the line-spread function resulting from the convolution of the different elements of the convolution process. That a Gaussian function forms a good approximation of the optical MTF of the eye, appears from a comparison of the high frequency behavior of the model with the measured data that will be given in section 3.9.

The quantity σ in Eq. (3.6) generally depends on the pupil diameter d of the eye lens. For very small pupil diameters, σ increases inversely proportionally with pupil size because of diffraction, and for large pupil diameters, σ increases about linearly with pupil size because of chromatic aberration and other aberrations. See Vos et al. (1976, Fig. 3). According to these authors, diffraction effects become noticeable only at pupil diameters smaller than 2 mm. Therefore, they may be neglected under normal viewing conditions. Therefore, it is assumed here that the dependence on pupil size can simply be expressed by the following equation:

$$\sigma = \sqrt{\sigma_0^2 + (C_{ab}d)^2} \qquad (3.7)$$

where σ_0 is a constant, C_{ab} is a constant that describes the increase of σ at increasing pupil size, and d is the diameter of the pupil. From an evaluation of contrast sensitivity measurements, it appears that for observers with good vision, σ_0 is about 0.5 arc min and C_{ab} is about 0.08 arc min/mm. The value of σ_0 is only partly determined by the optical effect of the eye lens. It is also determined by the density of the photo-receptors. As the density of the cones decreases with increasing distance from the center of the retina, σ_0 increases with this distance. See Chapter 4. However, for the normal situation of foveal vision treated in this chapter, σ_0 may be considered as constant.

The diameter d of the pupil generally depends on the average luminance of the observed object. To calculate the pupil size for a given luminance, the following simple approximation formula given by Le Grand (1969, p. 99) can be used:

$$d = 5 - 3\tanh(0.4\log L) \qquad (3.8)$$

where d is the pupil diameter in mm and L is the average luminance in cd/m^2. This expression is similar to other formulae, earlier given by Crawford (1936), Moon & Spencer (1944) and De Groot & Gebhard (1952). These formulae represent an average of various measurement data that show a large spread. Apart from the difference between different observers, this spread is also caused by the difference in the angular size of the object fields used in the experiments. Bouma (1965) investigated the effect of different field sizes. From his measurements an approximately quadratic dependence on field size can be derived. By assuming that Eq. (3.8) is valid for an average field size of $40°\times 40°$, one obtains the following approximation formula where also the field size is taken into account

$$d = 5 - 3 \tanh\{0.4 \log (LX_o^2/40^2)\} \tag{3.9}$$

where X_o is the angular field size of the object in degrees. For a rectangular field X_o^2 has to be replaced by $X_o Y_o$, and for a circular field X_o^2 has to be replaced by $\pi/4 \times D^2$ where D is the field diameter in degrees. This expression will generally be used here as a refinement of Eq. (3.8). It is in fact only valid for young adult observers. At older ages, the pupil size decreases with age. See, for instance, Kumninck (1954, Fig. 4) and Bouma (1965, Fig. 7.30).

3.4 Photon noise

The effect of photon noise on the contrast sensitivity of the eye was first discovered by de Vries (1943) and was later evaluated by Rose (1948) who explicitly cites the paper of de Vries. Often an earlier paper of Rose (1942) is cited for this effect, but this paper does not contain any mention of this effect.

According to de Vries the detection threshold at low luminance levels is determined by fluctuations in the number of photons that cause an excitation of the photo-receptors. Let the number of these photons within an area $\Delta x \Delta y$ and time Δt be n. For the statistical process of the arbitrary arriving photons, the standard deviation of this number is equal to $\sqrt{\bar{n}}$ where $\bar{n}$ is the average value of n. This average value may be expressed in the average flux density j of the photons with the equation

$$\bar{n} = j \Delta x \Delta y \Delta t \tag{3.10}$$

For the relative standard deviation σ_n of n holds

$$\sigma_n = \frac{\sqrt{\bar{n}}}{\bar{n}} = \frac{1}{\sqrt{j \Delta x \Delta y \Delta t}} \tag{3.11}$$

According to de Vries these fluctuations form the background noise that hampers the observation of an object. Application of Eq. (2.42) gives for the spectral density of the photon noise

$$\Phi_{ph} = \sigma_n^2 \Delta x \Delta y \Delta t \tag{3.12}$$

where σ_n has replaced σ_n and ϕ_{ph} has replaced ϕ_n. Inserting Eq. (3.11) in this expression gives

$$\Phi_{ph} = \frac{1}{j} \tag{3.13}$$

This equation says that the spectral density of photon noise is equal to the inverse of the average flux density of the photons on the retina that cause an excitation of the photo-receptors. The flux density on the retina can be derived from the luminous intensity of the light entering the eye with the following equation:

$$j = \eta p E \qquad\qquad (3.14)$$

where η is the quantum efficiency of the eye, p is the photon conversion factor for the conversion of light units in units for the flux density of the photons and E is a quantity that describes the retinal illuminance. Each of these quantities will be treated in more detail in this section.

The quantum efficiency η is defined here as the average number of photons causing an excitation of the photo-receptors, divided by the number of photons entering the eye. Although the quantum efficiency varies in principle with the wavelength, the wavelength dependence will be taken into account in the photon conversion factor. See Appendix A of this chapter. In this way η represents the quantum efficiency at the maximum of the spectral sensitivity curve. Contrary to what one would expect, the quantum efficiency of the eye is very low. From an evaluation of contrast sensitivity measurements, it appears that η is about 3% or less (See, for instance, Table 3.1 in section 3.9.15). Van Meeteren (1978) found even values of 2% and less by measuring the contrast sensitivity with and without artificial image intensification. He tried to explain the low quantum efficiency by various causes of losses. A part of the light is lost by absorption in the ocular media, another part falls in the interstices between the photo-receptors, a part of the light falling on a photo-receptor is not absorbed, and finally not every absorbed photon causes an excitation. However, van Meeteren could not explain the low quantum efficiency that he measured by an estimate of these losses. The low quantum efficiency might be explained by fluctuations in the excitation of the photo-receptors. If these fluctuations are not negligible, they form an additional noise source that can be translated in an effectively lower quantum efficiency.

The photon conversion factor p in Eq. (3.14) is defined as the number of photons per unit of time, per unit of angular area, and per unit of luminous flux per angular area of the light entering the eye. Absorption losses and other losses are already taken into account in the quantum efficiency η. The number of photons generally depends on the spectral wave length of the light. Equations for the calculation of the photon conversion factor from the spectral composition of the light source are given in Appendix A of this chapter. They are derived from basic photometric and physical quantities. For the calculation of the photon conversion factor a distinction has to be made between photopic vision (= daylight vision) where the cones act as photo-receptor, and scotopic vision (= night vision), where the rods act as photo-receptor. The spectral sensitivity for photopic vision is different from that for scotopic vision, as the cones are less sensitive for blue light and the rods are less sensitive for red light. In Table 3.2 of Appendix A of this chapter, numerical values of the photon conversion factor are given for different light sources. Although the use of the contrast sensitivity model given here is restricted to photopic viewing, data for scotopic viewing are also given as general information.

The quantity E in Eq. (3.14) is proportional to the retinal illuminance and can be calculated from the luminance L of the object and the pupil size d with the following equation

$$E = \frac{\pi d^2}{4} L \qquad (3.15)$$

If the pupil size is expressed in mm and the luminance in cd/m^2, E is given in Troland, indicated with Td. 1 Troland corresponds with a retinal illuminance of about 2×10^{-3} lux, taking into account the absorption of the light in the ocular media and the angular area of the pupil seen from the retina. Although the Troland does not have the dimension of illuminance, it is for practical reasons chosen as a measure of retinal illuminance. The transition between scotopic vision and photopic vision occurs at a level between 1 and 10 Td. The pupil size can be measured, or can be derived from the luminance with Eq. (3.9).

For the photopic viewing conditions used here, Eq. (3.15) has to be corrected for the Stiles-Crawford effect. For light falling on the cones, Stiles & Crawford (1933) found that rays entering near the edge of the pupil are visually much less effective than rays near the center of the pupil. From the work by Stiles and Crawford, Moon & Spencer (1944) and Jacobs (1944) derived an expression that forms a modification of Eq. (3.15) and may be written in the following form:

$$E = \frac{\pi d^2}{4} L \{ 1 - (d/9.7)^2 + (d/12.4)^4 \} \qquad (3.16)$$

where d is expressed in mm. This expression will be used in the model. Although the decrease of the quantum efficiency by the Stiles-Crawford effect could also have been taken into account in the quantum efficiency η, the use of this expression is preferred here for practical reasons. For large pupil sizes, the correction for the Stiles-Crawford effect can amount to 50%.

By combining Eqs. (3.13) and (3.14) one obtains

$$\Phi_{ph} = \frac{1}{\eta p E} \qquad (3.17)$$

According to this equation and Eq. (3.5), contrast sensitivity increases at low luminance levels with the square root of retinal illuminance. At these levels the effect of photon noise is so large that the effect of neural noise may be neglected. This square root behavior is known as *de Vries-Rose law*. An example of this behavior will be shown by the measurement data given in Fig. 3.21 of section 3.9.13.

3.5 Neural noise

In the model, it is assumed that neural noise is caused by statistical fluctuations in the signal transported to the brain. Contrary to electronic image systems, where usually only one wire is used for the transport of a signal, the image formed on the retina of the eye is transported to the brain by many fibers in parallel. When the image consists of a uniformly illuminated field, one may not expect that the different parts of this field will be reproduced by all nerve fibers in the same amount. Small differences between the different fibers will cause noise in the image arriving in the brain. The size of these differences can be estimated from the spectral density of the noise. From a comparison of contrast sensitivity measurements with the results obtained with the model, the spectral density Φ_0 of the neural noise may be estimated to be about 0.03×10^{-6} sec $\deg^2$ (This follows, for instance, from the measurements shown in sections 3.9.11 and 3.9.12). From Eq. (2.42) follows for the relative standard deviation of the signal transported by an individual nerve fiber:

$$\sigma = \sqrt{\frac{\Phi_0}{\Delta x\, \Delta y\, \Delta t}} \qquad\qquad (3.18)$$

where $\Delta x \Delta y$ is the retinal angular area covered by one nerve fiber, and Δt is the integration time of the visual system. The density of ganglion cells from which the nerve fibers originate may be estimated to be about 1,800 cells per $\deg^2$ in the center of the retina (See section 4.2 of Chapter 4). This means that $1/(\Delta x \Delta y) \approx 1,800/\deg^2$. If for the integration time of the eye a value of 0.1 sec is used, the relative standard deviation of the signal transported by the individual nerve fibres becomes

$$\sigma = \sqrt{\frac{0.03 \cdot 10^{-6} \cdot 1,800}{0.1}} = 0.023$$

This is a fluctuation of 2.3%, which may be considered as a reasonable value.

In the model, it is assumed that neural noise does not depend on retinal illuminance. At high retinal illuminance levels where the effect of photon noise decreases, neural noise remains as only noise source. According to Eq. (3.5) contrast sensitivity then becomes independent of luminance. This behavior is known as *Weber's law*. An example of this behavior will be shown by the measurement data given in Fig. 3.21 of section 3.9.13.

3.6 Lateral inhibition

In our model, it is assumed that the luminance signal and the added photon noise are filtered in the neural system by a lateral inhibition process that attenuates low spatial frequency components. Since the contrast sensitivity appears to decrease linearly with the inverse of spatial frequency at low spatial frequencies, the effect of lateral inhibition can be characterized by an MTF that increases linearly with spatial frequency at low spatial frequencies up to 1 at a certain spatial frequency and remains further constant at higher spatial frequencies. From an investigation of natural scenes, Field (1987) found that the amplitude of the spatial frequency components of natural images decreases linearly with spatial frequency. This property of natural scenes is obviously compensated at low spatial frequencies by the increase of the MTF in this area due to the lateral inhibition. The existence of lateral inhibition may, therefore, probably be explained by the fact that the eye can make in this way a more efficient use of the dynamic range of signals that it can handle.

As was already supposed by Schade (1956) and was experimentally confirmed by Enroth-Cugell & Robson (1966) in their investigation with cats, lateral inhibition consists of the subtraction of a spatially lowpass filtered signal from a signal that is directly collected from the photo-receptors. Enroth-Cugell and Robson described the point-spread function of this process by a difference of two Gaussian functions, which has the form of a Mexican hat. This model is usually called *DOG model* (difference of Gaussians). However, it leads to a quadratic increase of contrast sensitivity at low spatial frequencies, whereas measurements of the contrast sensitivity clearly show a linear increase. Therefore, a different approach will be followed here.

From an evaluation of published contrast sensitivity measurements, we found that the MTF of the lateral inhibition process can well be described by the following approximation formula (Barten, 1992):

$$M_{lat}(u) = \sqrt{1 - e^{-(u/u_0)^2}} \qquad (3.19)$$

This function is shown by the solid curve in Fig. 3.2. It gives a linear increase of the MTF with spatial frequency up to a value 1 at a spatial frequency u_0 above which the lateral inhibition ceases. From a best fit of the model with the published contrast sensitivity measurements given in section 3.9, it appears that u_0 is about 7 cycles/deg. As contrast sensitivity is nearly independent of orientation, certainly at low spatial frequencies, it may further be assumed that the lateral inhibition process is rotationally symmetric.

As the MTF of the lateral inhibition process is the result of the subtraction of a lowpass filtered signal from a signal that is directly obtained from the photo-receptors, the MTF of the lateral inhibition process may also be described by

$$M_{lat}(u) = 1 - F(u) \qquad (3.20)$$

where $F(u)$ is the MTF of the spatial lowpass filter. Combination of Eqs. (3.19) and (3.20) gives

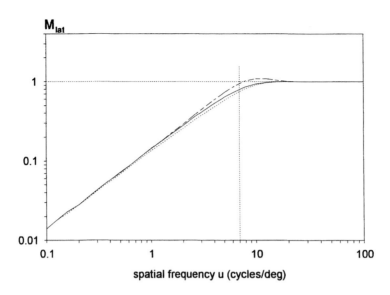

spatial frequency u (cycles/deg)

Figure 3.2: Solid curve: MTF of the lateral inhibition process given by Eq. (3.19) with $u_0 = 7$ cycles/deg. Dotted curve: MTF calculated with Eqs. (3.20) and (3.21) for the receptive field given by Eq. (3.23). Dashed curve: MTF calculated with Eqs. (3.20) and (3.25) for the annular receptive field given by Eq. (3.24).

$$F(u) = 1 - \sqrt{1 - e^{-(u/u_0)^2}} \qquad (3.21)$$

The point-spread function that gives such an MTF can be found by an inverse Hankel transform of this expression. See, for instance, Papoulis (1968, pp. 140-145). The result can be numerically calculated but cannot be represented in mathematical form. This becomes, however, possible, if Eq. (3.21) is replaced by the following expression:

$$F(u) = 0.5 e^{-2u/u_0} + 0.5 e^{-(u/u_0)^2} \qquad (3.22)$$

The MTF given by this function has nearly the same shape as the MF given by Eq. (3.19). It is shown by the dotted curve in Fig. 3.2. An inverse Hankel transformation of this function gives

$$f(r) = \frac{0.25 \pi u_0^2}{(1 + \pi^2 u_0^2 r^2)^{3/2}} + 0.5 \pi u_0^2 e^{-\pi^2 u_0^2 r^2} \qquad (3.23)$$

This function describes the *receptive field* of the inhibition process.

After the classical DOG model for the lateral inhibition process, a model consisting of a ring of Gaussians has been introduced. See, for instance, Young (1991). This model is called *DOOG model* (difference of offset Gaussians). These Gaussians form together an annular shaped lowpass filter, instead of the continuous Gaussian lowpass filter used in the DOG model. An annular lowpass filter seems to give a better description of the lateral inhibition process. The lowpass filter given by

Eq. (3.23) can be changed in an annular filter by modifying Eq. (3.23) into

$$f(r) = \frac{0.25 \pi u_0^2}{(1 + \pi^2 u_0^2 r^2)^{3/2}} + 1.5 \pi u_0^2 e^{-\pi^2 u_0^2 r^2} - 1.75 \pi u_0^2 e^{-1.75 \pi^2 u_0^2 r^2} \qquad (3.24)$$

A Hankel transform of this expression gives

$$F(u) = 0.5 e^{-2u/u_0} + 1.5 e^{-(u/u_0)^2} - 1.0 e^{-\frac{1}{1.75}(u/u_0)^2} \qquad (3.25)$$

This function gives a slightly different description of the MTF of the lateral inhibition process than Eq. (3.19). The MTF derived from this function is shown by the dashed curve in Fig. 3.2.

Fig. 3.3 shows a cross-section of the total point-spread function of the eye obtained by a combination of the optical point-spread function of the eye with the point-spread function of the annular lowpass filter given by Eq. (3.24). The shape of the annular lowpass filter is shown by the dotted curve in this figure, which is plotted with a negative sign to indicate the subtraction made by this filter. For u_0 the mentioned value of 7 cycles/deg is used. The figure further shows measurement data of the total point-spread function given by Blommaert et al. (1987). These data were obtained with a sophisticated perturbation technique based on peak detection of a combination of sub-threshold stimuli. The measurements were made with an artificial

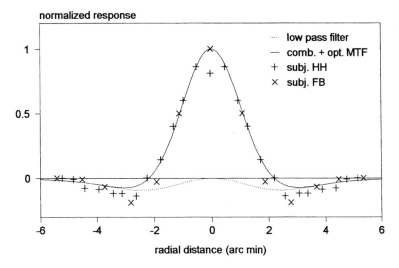

Figure 3.3: Solid curve: cross-section of the total point-spread function of the eye obtained by a combination of the optical point-spread function of the eye with the annular point-spread function of the low-pass inhibition filter given by Eq. (3.23). Dotted curve: cross-section of the annular point-spread function of the lowpass inhibition filter. Data points: measurements of the total point-spread function by Blommaert et al. (1987). For the calculation of the solid curve, the σ of the optical point-spread function has been adapted to the measurements.

pupil of 2 mm and a retinal illuminance of 1200 Td. The value used for σ in the calculated point-spread function has been adapted to the measurements and appears to be somewhat higher than the usual value of 0.5 arc min. Apart from this, the calculations reasonably agree with the measurements. However, the measurements show slightly deeper negative side lobes.

Although the annular filter might give a somewhat better description of the receptive field of the lateral inhibition process, still some uncertainties remain. Therefore, and for the sake of simplicity, still the simple formula given by Eq. (3.19) will be used in the model given here.

3.7 Monocular vision versus binocular vision

In comparing visual thresholds, it is important to take into account whether the observation is made with one eye, or with both eyes. At binocular vision, the information of both eyes is combined, while the internal noise of both eyes is not correlated, as the noise is separately generated in each eye. This can be considered as a doubling of the effective integration area. According to Eq. (2.43), the modulation of the internal noise is then reduced with a factor $\sqrt{2}$. So, the contrast sensitivity for binocular viewing increases with a factor $\sqrt{2}$ compared with monocular viewing. This holds only if the information of both eyes is completely combined, and if there is no noise added to the combined information processed in the brain. From measurements, it appears that this is indeed the case. Campbell & Green (1965) found that the contrast sensitivity for binocular viewing is a factor $\sqrt{2}$ higher than for monocular viewing and van Meeteren (1973) later also found the same results.

As binocular vision is the most common type of viewing, the factor $\sqrt{2}$ is used as standard in the contrast sensitivity model given here. The contrast sensitivity given by Eq. (3.5) has, therefore, to be multiplied with this factor. For monocular vision the contrast sensitivity is a factor $\sqrt{2}$ smaller. If the contrast sensitivity is limited by external noise, the noise presented to both eyes is correlated. Then the contrast sensitivity has also to be taken a factor $\sqrt{2}$ smaller. In this situation it makes no difference if the object is observed with one eye or with two eyes.

3.8 Complete model

After correcting Eq. (3.5) with a factor $\sqrt{2}$ for binocular viewing and after inserting Eq. (2.51) given in Chapter 2 and the equations given in the preceding sections, the following formula for the spatial contrast sensitivity function at binocular vision is obtained:

$$S(u) = \frac{1}{m_t(u)} = \frac{M_{opt}(u)/k}{\sqrt{\frac{2}{T}\left(\frac{1}{X_o^2} + \frac{1}{X_{max}^2} + \frac{u^2}{N_{max}^2}\right)\left(\frac{1}{\eta p E} + \frac{\Phi_0}{1 - e^{-(u/u_0)^2}}\right)}} \qquad (3.26)$$

For monocular vision, $S(u)$ is a factor $\sqrt{2}$ smaller. This means that the factor 2 under the square root sign has to be replaced by 4. In this equation, $M_{opt}(u)$ is the optical MTF given by Eq. (3.6), u is the spatial frequency, k is the signal-to-noise ratio, T is the integration time of the eye, X_o is the angular size of the object, X_{max} is the maximum angular size of the integration area, N_{max} is the maximum number of cycles over which the eye can integrate the information, η is the quantum efficiency of the eye, p is the photon conversion factor that depends on the light source and is given in Table 3.2 in Appendix A of this chapter, E is the retinal illuminance in Troland, Φ_0 is the spectral density of the neural noise, and u_0 is the spatial frequency above which the lateral inhibition ceases. This formula holds for the situation that the object dimensions in x and y directions are equal. For nonequal dimensions, the factor between the brackets that contains the object size has to be replaced by $1/XY$ where X and Y are given by Eqs. (2.48) and (2.49), respectively. The constants in the model have the following typical values:

k	= 3.0	T	= 0.1 sec	η	= 0.03
σ_0	= 0.5 arc min	X_{max}	= 12°	Φ_0	= 3×10^{-8} sec deg^2
C_{ab}	= 0.08 arc min/mm	N_{max}	= 15 cycles	u_0	= 7 cycles/deg

For T, it is assumed that the presentation time is long with respect to the integration time of the eye; otherwise Eq. (2.45) has to be used. The given constants are valid for an average observer, foveal vision and photopic viewing conditions. They have been obtained from a best fit with measurement data. For an arbitrary individual subject, only the values of σ_0, η, and k have to be adapted.

Fig. 3.4 shows the cumulative effect of various factors on the shape of the contrast sensitivity function. The figure has been calculated with Eq. (3.26) for a field size of $10°\times10°$ using the given typical values of the constants. The horizontal line at the top of the figure shows the ultimate limit of the contrast sensitivity for this field size. This limit is determined by neural noise. Lateral inhibition causes a linear attenuation of this limit at low spatial frequencies. The maximum number of cycles causes a decay at high spatial frequencies, which is further enforced by the optical MTF of the eye. Photon noise causes a further decrease of the contrast sensitivity and a change in shape of the contrast sensitivity function at lower luminance levels. The figure shows that for low luminance and not too low spatial frequency, the contrast

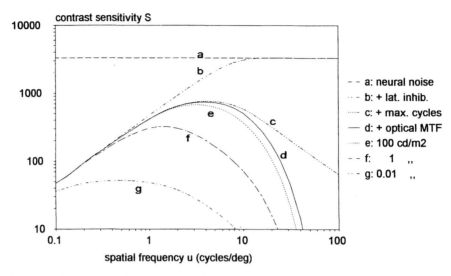

Figure 3.4: Cumulative effect of different factors on contrast sensitivity, calculated with Eq. (3.26) for a field size of $10° \times 10°$: (a) neural noise; (b) + lateral inhibition; (c) + limited number of cycles; (d) + optical MTF; (e), (f), and (g) + photon noise at 100 cd/m², 1 cd/m², and 0.01 cd/m², respectively.

sensitivity increases with the square root of the luminance, according to the de Vries-Rose law. The figure also shows that for high luminance or low spatial frequency, the contrast sensitivity is nearly independent of the luminance, according to Weber's law. The dependence of contrast sensitivity on field size is not shown in the figure, but will later be shown in Figs. 3.19 and 3.22 where the model is compared with contrast sensitivity measurements for different field sizes. These figures show that the field size causes a vertical shift of the low frequency part of the curves, whereas the high frequency part remains the same, due to the effect of the limited number of cycles.

3.9 Comparison with measurements

The contrast sensitivity function that can be calculated with the model will now be compared with several published measurements. For the directional orientation of the sinusoidal test patterns mentioned in these publications, it should be noted that "vertically oriented" means that the bars are vertically oriented and that the sinusoidal luminance variation takes place in horizontal direction. The measurements will be given in chronological order of publication. The constants σ_0, η, and k will in each case be adapted to obtain a best fit with the measurements. This fit will be made by trial and error. The constant k appears to influence the fit mainly at low spatial frequencies, η appears to influence mainly the fit at medium spatial frequencies, and σ_0 appears to influence mainly the fit at high spatial frequencies. It should be noted that for measurements with a series of data curves, a simultaneous fit will be made for all curves. If the contrast sensitivity is determined with the aid of a 2AFC method where the results do not correspond with 75% correct response, k will be corrected with Eq. (2.14) given in Chapter 2. A survey of the values used for σ_0, η, and k will be given in Table 3.1 at the end of this section.

3.9.1 Measurements by DePalma and Lowry

DePalma & Lowry (1962) measured the contrast sensitivity function at two different luminance levels: 1028 cd/m^2 and 69cd/m^2 (300 ftL and 20 ftL). The test object was a vertically oriented sinusoidal grating illuminated with a variable luminance and combined with a veiling illumination to obtain a variable contrast. The color temperature of the illumination was 2850 K. The measurements were made at a viewing distance of 35 inches (0.89 m) with a field size of 6°×6°. The observer looked at the test object with both eyes and without an artificial pupil. The modulation threshold was determined by the method of adjustment.

Measurements and calculations are shown in Fig. 3.5. The values of σ_0, η, and k used for the calculation were 0.45 arc min, 0.5%, and 3.0, respectively. The value of η is rather low. The general agreement between measurements and calculations is good, although the measured data merge somewhat earlier at low spatial frequencies than the calculated curves.

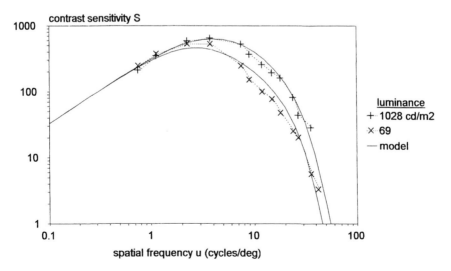

Figure 3.5: Contrast sensitivity function measured by DePalma & Lowry (1962) at two different luminance levels. Field size 6°×6°. Binocular viewing with a natural pupil. The solid curves have been calculated with Eq. (3.26).

3.9.2 Measurements by Patel

Patel (1966) measured the contrast sensitivity function at four different retinal illuminance levels ranging from 3 Td to 1000 Td. The test object was a vertically oriented sinusoidal grating pattern generated on the screen of an oscilloscope tube provided with a green phosphor (P31). The luminance of the pattern was adjusted by using appropriate filters. The measurements were made at a distance of 1 m with a field size of 2°×2°. The observer looked at the test object with one eye through an artificial pupil of 2 mm. The modulation threshold was determined by the method of adjustment. Measured data were given for only one subject, a male student between 20 and 25 years of age, whose measurement results were considered "typical."

Measurements and calculations are shown in Fig. 3.6. The values of σ_0, η, and k used for the calculation were 0.5 arc min, 3%, and 3.1, respectively. The measurements at 10, 100, and 1000 Td generally agree with the calculations, apart from a severe dip at about 2 cycles/deg. Such a dip is not found in measurements by other authors, so that it has to be assumed that it was caused by some particular measurement error. This possibility was already mentioned by the authors. Another difference between measurements and calculations is the fact that the measured data for 3 Td do not correspond with calculations for 3 Td, but with calculations for 0.3 Td shown by the dotted curve in the figure. This is probably also due to some error.

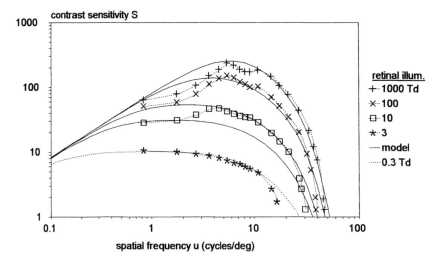

Figure 3.6: Contrast sensitivity function measured by Patel (1966) at four different retinal illuminance levels. Field size 2°×2°. Monocular viewing with an artificial pupil of 2 mm. The solid curves have been calculated with Eq. (3.26) for the retinal illuminance of the data. The dotted curve gives a fit with the data for 3 T, but has been calculated for 0.3 Td.

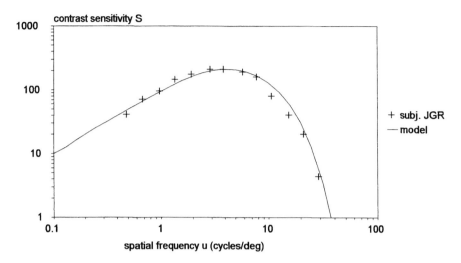

Figure 3.7: Contrast sensitivity function measured by Robson (1966) at a luminance of 20 cd/m². Field size 2.5°×2.5°. Binocular viewing with a natural pupil. The curve through the data points has been calculated with Eq. (3.26).

3.9.3 Measurements by Robson

Robson (1966) measured the contrast sensitivity function at a luminance of 20 cd/m². The test object was a vertically oriented sinusoidal grating pattern generated on the screen of an oscilloscope tube provided with a green phosphor (P31). Although the grating pattern was simultaneously modulated with a temporal frequency of 1 Hz, this frequency may be assumed to be sufficiently low to consider the measurements as static. See Chapter 5. The angular size of the test object was 2.5°×2.5° and the measurements were made at a viewing distance of 1 m. The observer looked at the test object with both eyes and without an artificial pupil. The modulation threshold was probably determined by the method of adjustment. The author himself (a corrected myope) was the observer.

Measurements and calculations are shown in Fig. 3.7. The values of σ_0, η, and k used for the calculation were 0.53 arc min, 2.0%, and 4.5, respectively. The measurements show a very good agreement with the calculations.

3.9.4 Measurements by van Nes and Bouman

Van Nes & Bouman (1967)made similar measurements as Patel, but with monochromatic light and with a larger field size. Three types of monochromatic light were used: green light with a wavelength of 525 nm, red light with a wavelength of 650 nm, and blue light with a wavelength of 450 nm. The measurements with green light extended over a retinal illuminance range of six decades. The test object was a vertically oriented transparent sinusoidal grating illuminated by a variable luminance and combined with a veiling luminance to obtain a variable contrast. The angular size of the test object was 4.5° in horizontal direction and 8.25° in vertical direction. The surrounding field was completely dark. The observer looked at the test object with one eye through an optical system with an artificial pupil of 2 mm. The modulation threshold was determined by the method of adjustment where the observer could vary the modulation in steps of one tenth of a decade. In this procedure a lower and a higher limit were determined of which the arithmetic is used here for the measurement data. The lower and higher limits differed each about 12% from the average. See van Nes (1968). The first author was the observer.

Measurements and calculations for green, red, and blue light are shown in Fig.3.8, Fig. 3.9, and Fig. 3.10, respectively. It should be remarked that the light level of the lowest curves in these figures is actually scotopic, whereas the model is only valid for photopic conditions. For all three colors the same values for η and k could be used in the calculations 30% and 2.7, respectively, whereas for σ_0 different values had to be used: 0.45 arc min for green light, 0.54 arc min for red light, and 0.50 arc min for blue light. For blue light, a photon conversion factor was used that was derived from the CIE $V_{10}(\lambda)$ curve for fields of 10° and larger (See Appendix A of this chapter), because of the large field size of the measurements. This factor differs from the factor mentioned in Table 3.2, which is based on the commonly used $V_2(\lambda)$ curve

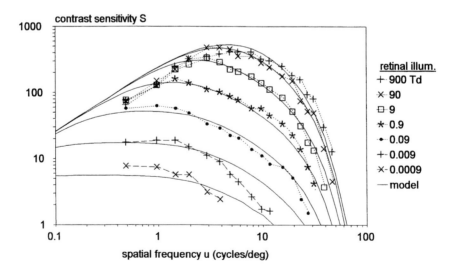

Figure 3.8: Contrast sensitivity function measured by van Nes & Bouman (1967) over a large range of retinal illuminance levels using monochromatic green light with a wavelength of 525 nm. Field size 4.5°×8.25°. Monocular viewing with an artificial pupil of 2 mm. The solid curves have been calculated with Eq. (3.26).

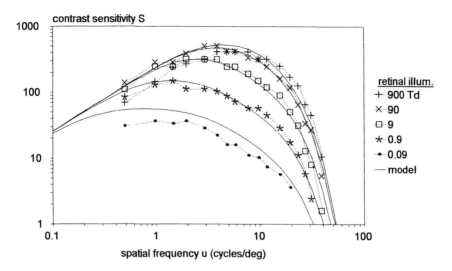

Figure 3.9: Same as Fig. 3.8, but for monochromatic red light with a wavelength of 650 nm.

that is only valid for small fields. For the other colors, the field size makes no difference. The value of 30% for the quantum efficiency η of all three colors is unlikely high compared with the results of other investigations. For the more usual value of 3.0%, the measured retinal illuminance should have been a factor 10 higher.

Measurements and calculations further show a good agreement over a large

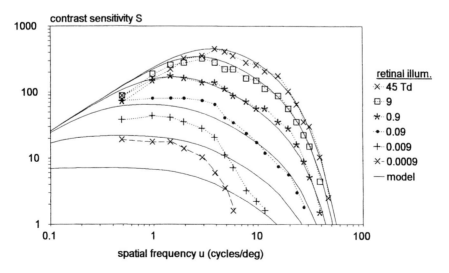

Figure 3.10: Same as Fig. 3.8, but for monochromatic blue light with a wavelength of 450 nm.

range of retinal illuminance levels. Not only the vertical position of the calculated curves agrees with the measurements, but also their changing shape at variation of the retinal illuminance. An exception is formed by the lowest curves of which the light level is scotopic, instead of photopic, as assumed in the model. The measurement data for the lowest curves in Figs. 3.9 and 3.10 clearly show the low sensitivity of the rods for red light and the high sensitivity of the rods for blue light, respectively, compared with the sensitivity of the cones used in the calculation. The measurement data for the lowest curves in Figs. 3.8 and 3.10 further show that scotopic resolution is much less than photopic resolution.

3.9.5 Measurements by Campbell and Robson

Campbell & Robson (1968) measured the contrast sensitivity function at a luminance of 500 cd/m^2. The test object was a vertically oriented sinusoidal grating pattern generated on the screen of a monochrome CRT provided with a white phosphor (P4). The modulating voltage was switched on and off at a rate of 0.5 Hz. This rate may be assumed to be still sufficiently low to consider the presentation as static (See Chapter 5). Measurements for low spatial frequencies were made at a viewing distance of 0.57 m with a field size of 10°×10°, whereas measurements for high spatial frequencies were made at a distance of 2.85 m with a field size of 2°×2°. The observer looked at the test object with one eye through an artificial pupil of 2.5 mm. The modulation threshold was determined by the method of adjustment where

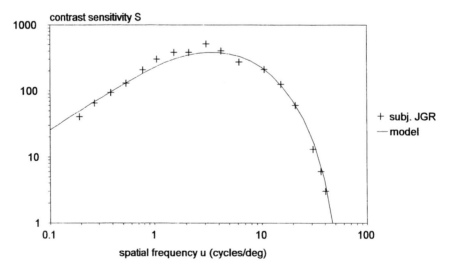

Figure 3.11: Contrast sensitivity function measured by Campbell & Robson (1968) at a luminance of 500 cd/m^2. Field size 10°×10°. Monocular viewing with an artificial pupil of 2.5 mm. Subject JGR. The curve through the data points has been calculated with Eq.(3.26).

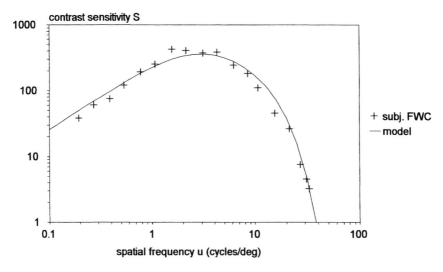

Figure 3.12: Same as Fig. 3.11, but for subject FWC.

the observer varied the modulation until the grating was barely detectable. Either five or ten observations were made to determine each threshold. Both authors served as subjects.

Measurements and calculations for the second author are shown in Fig. 3.11. The values of σ_0, η, and k used for the calculation were 0.53 arc min, 2.5%, and 3.9, respectively. For the sake of simplicity all calculations have been made for a field size of $10° \times 10°$, also for the six highest spatial frequencies measured with a field size of $2° \times 2°$. For these frequencies this hardly makes any difference. As can be seen from the figure, the measurements show a very good agreement with the calculations. The observer was the same person as the observer at the measurements given in section 3.9.3 where for this subject the same value was found for σ_0 and nearly the same value for η. These measurements were made under different conditions of luminance and field size and were made with both eyes and a natural pupil.

Measurements and calculations for the first author are shown in Fig. 3.12. The value of k for this observer was the same, but the values of σ_0 and η were 0.69 arc min and 1.0%, respectively. The higher value of σ_0 and the lower value of η could be caused by the higher age of this subject.

3.9.6 Measurements by Watanabe et al.

Watanabe et al. (1968) measured the contrast sensitivity function at a luminance of 34 cd/m^2 (10 ftL). The test object was a vertically oriented sinusoidal grating pattern

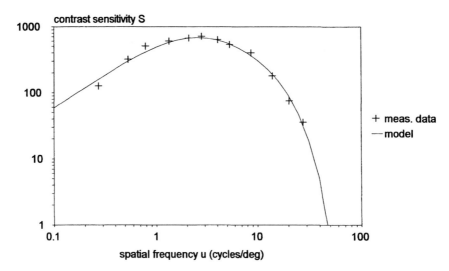

Figure 3.13: Contrast sensitivity function measured by Watanabe et al. (1968) at a luminance of 34 cd/m². Field size 19°×14°. Binocular viewing with a natural pupil. The curve through the data points has been calculated with Eq. (3.26).

generated on the screen of a monochrome TV monitor provided with a white phosphor (P4). The size of the test object was 24 cm in horizontal direction and 18 cm in vertical direction. Measurements for low spatial frequencies were made at a viewing distance of 0.72 m, whereas measurements for high spatial frequencies were made at a viewing distance of 3.24 m. This corresponded with a field size of 19°×14° and 4°×3°, respectively. The observer looked at the test object with both eyes and without an artificial pupil. The modulation threshold was determined by the method of adjustment where the observer varied the modulation to the point where the grating pattern just disappeared. The reported data are from one of the subjects that took part in the experiments.

Measurements and calculations are shown in Fig. 3.13. The values of σ_0, η, and k used for the calculation were 0.48 arc min, 4%, and 3.0, respectively. As with the measurements by Campbell and Robson, the calculations have been made for only the largest field size. Also in this case, this hardly makes any difference for the high spatial frequencies. Both types of measurements show a very good agreement with the calculations.

3.9.7 Measurements by Sachs et al.

Sachs et al. (1971) measured the contrast sensitivity for five different spatial frequencies (1.4, 2.8, 5.6, 11.2 and 22.4 cycles/deg) at a luminance of 64 cd/m² (20 mL). The test object was a vertically oriented sinusoidal grating pattern generated on the screen of a monochrome CRT provided with P31 phosphor. The test object was surrounded by a large sheet of cardboard at the same luminance as the stimulus. The measurements were made at a viewing distance of 2.4 m with a field size of 4.5°×4.5°. The observer looked at the test object with both eyes and without an artificial pupil. The observer was MS, the first author.

Besides other data, measurements were also given of the psychometric function. From these data not only the contrast sensitivity could be determined, but also the k value that occurred in the measurements. This is different from other published measurements given in this chapter, where only data of the contrast sensitivity were available and the k value had to be determined by a best fit with the measurements.

Fig. 3.14 shows the psychometric functions for the five spatial frequencies, plotted in a normalized way as a function of m/m_t, as described in Chapter 2. The curve through the data points has been calculated with Eqs. (2.2) through (2.4) for the combined results of the normalized data. For this curve, the k value appeared to be 3.73. Fig 3.15 shows the contrast sensitivity obtained from the psychometric functions of these data as a function of the spatial frequency, together with the contrast sensitivity function calculated with the k value obtained from the psychometric function. The values of σ_0 and η used for this calculation were 0.59 arc min and 3%, respectively. The agreement between measurements and calculations shows

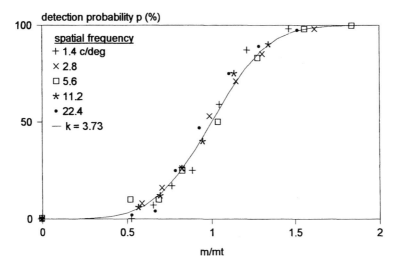

Figure 3.14: Normalized psychometric function for measurements by Sachs et al. (1971) at five different spatial frequencies. The curve through the data points has been calculated with Eqs. (2.2) through (2.4) for the combined results of the normalized data. For this curve, $k = 3.73$.

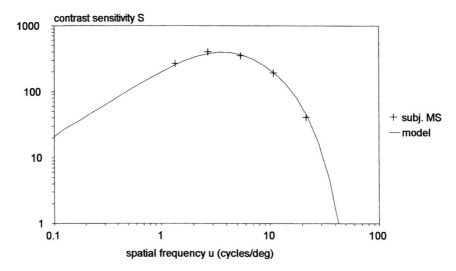

Figure 3.15: Contrast sensitivity function determined from measurements of the psychometric function by Sachs et al. (1971). Luminance 64 cd/m². Field size 4.5°×4.5°. Binocular viewing with a natural pupil. The curve through the data points has been calculated with Eq. (3.26) using for k the value determined from the psychometric function for the combined set of data.

that the values of u_0 and Φ_0 used in Eq. (3.26) are in good agreement with the measurements, as otherwise a different k value had to be used to obtain a fit with the data.

3.9.8 Measurements by van Meeteren and Vos

Van Meeteren & Vos (1972) made measurements at various luminance levels similarly as van Nes & Bouman. However, they used white light, both eyes and a natural pupil, in order to study contrast sensitivity under conditions that were closer to natural vision. The measurements extended over a luminance range of five decades from 10^{-4} cd/m² to 10 cd/m². The test objects consisted of slides with vertically and horizontally oriented sinusoidal grating patterns projected on a white screen at a distance of 3.5 m from the observer. The field size was 17° in horizontal direction and 11° in vertical direction. A uniform luminance was superimposed on this field by a second projector. Modulation and luminance level were partly varied by inserting neutral density filters and partly varied by controlling the lamp currents. The color temperature of the light source was 2850 K. The observer looked at the test object with both eyes and without an artificial pupil. Horizontal and vertical gratings were presented in random order to the subject, who had to say "horizontal," "vertical," or "no choice." The transition point from "no choice" answers to correct answers was used as threshold. This threshold appeared to correspond with 75% correct response

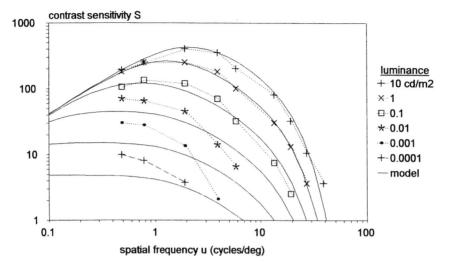

Figure 3.16: Contrast sensitivity function measured by van Meeteren & Vos (1972) over a luminance range of five decades. Field size 17°×11°. Binocular viewing with a natural pupil. The solid curves have been calculated with Eq. (3.26).

in a two-alternative forced-choice experiment. The measurements were made by two subjects. The results were averaged over both subjects and both pattern orientations.

Measurements and calculations are shown in Fig. 3.16. The values of σ_0, η, and k used for the calculation were 0.5 arc min, 3%, and 4.0, respectively. The agreement between measurements and calculations for the three highest luminance levels is good. Deviations from the curves for the three lowest luminance levels are caused by the fact that the light level of these curves is scotopic, whereas photopic conditions were assumed in the model.

3.9.9 Measurements by Howell and Hess

Howell & Hess (1978) measured the contrast sensitivity function at a luminance of 100 cd/m². The test object was a vertically oriented sinusoidal grating pattern generated on the screen of a television tube provided with a white phosphor (P4). The surrounding area was matched to the stimulus with respect to luminance and color. The test object had a width of 5 cycles and a height of 20 cycles. In this way the authors obtained a constant number of cycles for the integration area at all spatial frequencies. The spatial frequency was varied by varying the size of the object and by varying the viewing distance. By doing so, the viewing distance varied between 0.23 m and 5.7 m. The observer looked at the test object with both eyes and without an artificial pupil. The modulation threshold was determined by the method of adjust-

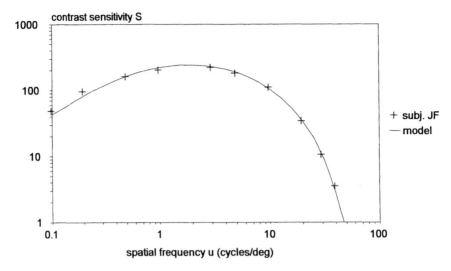

Figure 3.17: Contrast sensitivity function measured by Howell & Hess (1978) at a luminance of 100 cd/m^2. The test field had a width of 5 cycles and a height of 20 cycles. Binocular viewing with a natural pupil. The curve through the data points has been calculated with Eq. (3.26). The slightly flattened shape is caused by the use of a fixed number of cycles for all spatial frequencies, instead of a fixed field size.

ment where the observer could vary the modulation in steps of 0.5%. The data are the averages of at least five measurements. Two subjects took part in this experiment. Both subjects were corrected myopes. The data of one subject, JF, are used here.

Measurements and calculations are shown in Fig. 3.17. The values of σ_0, η, and k used for the calculation were 0.47 arc min, 3%, and 5.0, respectively. The measurements show a very good agreement with the calculations. The flattened shape of the curve compared with other measurements is caused by the fact that a fixed number of cycles was used in the experiment, instead of a fixed field size. This reduces the angular field size at increasing spatial frequency. The good agreement between measurements and calculations shows that this aspect is also well taken into account in the model.

3.9.10 Measurements by Virsu and Rovamo

Virsu & Rovamo (1979) measured the contrast sensitivity as a function of the number of cycles at different spatial frequencies. The luminance was 10 cd/m^2. The test object was a vertically oriented sinusoidal grating pattern generated on the screen of a high resolution monitor provided with a white phosphor (P4). The shape of the grating patterns was square, except for a few measurements below 1 cycle. In these situations one full cycle was used in horizontal direction, but the height in the other

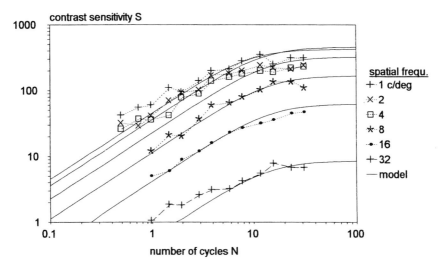

Figure 3.18: Contrast sensitivity as a function of the number of cycles measured by Virsu & Rovamo (1979) at a luminance of 10 cd/m². Binocular viewing with a natural pupil. The solid curves have been calculated with Eq. (3.26).

direction was smaller. The luminance of the surrounding area was the same as that of the test object. Viewing distance was varied in addition to a variation of the size of the grating, to achieve a large variation in angular size of the object at all spatial frequencies. The observer looked at the test object with both eyes and without an artificial pupil. The modulation threshold was determined with a 2AFC method where the threshold corresponded with 84% correct response. Gratings with zero and non-zero modulation were presented in random order. Seven subjects with ages between 25 and 36 participated in the experiments. The reported data are from one subject, the second author.

Measurements and calculations are shown in Fig. 3.18. In this figure contrast sensitivity is plotted as a function of the number of cycles, with spatial frequency as parameter. The values of σ_0, η, and k used for the calculation were 0.46 arc min, 2.5%, and 3.2, respectively, after correcting k for the difference between 84% and 75% correct response with the aid of Eq. (2.14). Apart from the lowest spatial frequencies the calculated curves agree well with the measurements. Measurements and calculations show a saturation of the contrast sensitivity at about 15 cycles.

3.9.11 Measurements by Carlson

Carlson (1982) measured the contrast sensitivity function for a large range of field sizes extending from 0.5° to 60°. The luminance of the test object was 108 cd/m²

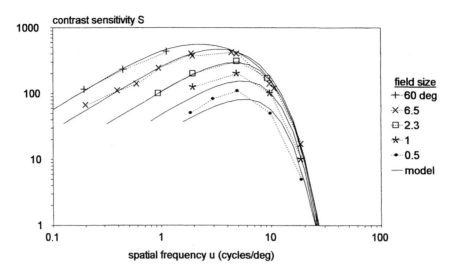

Figure 3.19: Contrast sensitivity function measured by Carlson (1982) at a luminance of 108 cd/m² for square test fields with a large range of angular sizes. Binocular viewing with a natural pupil. The solid curves have been calculated with Eq. (3.26).

(34 mL) with a surrounding luminance of one tenth of this value. The test object was a vertically oriented sinusoidal grating pattern with a square size. Patterns with angular field sizes of 0.5°, 1.0°, 2.3°, and 6.5° were generated on the screen of a monochrome television monitor, whereas patterns with angular field sizes of 6.5° and 60° were projected on the screen of an optical projection system. In both situations the viewing distance was 1.9 m. Both types of measurements gave the same results at an angular size of 6.5°. The observer looked at the test object with both eyes and without an artificial pupil. The modulation threshold was determined by the method of adjustment. The measurements were made with two subjects, one of which was the author. Ten readings were taken at each measurement point for each observer, and the results were averaged.

Measurements and calculations are shown in Fig. 3.19. The values of σ_0, η, and k used for the calculation were 1.1 arc min, 3.0%, and 3.8, respectively. The value of σ_0 is very high compared with other measurements. It was probably caused by the fact that one of the two subjects had to use very strong glasses. The measurements show a good agreement with the calculations. They further show a good illustration for the dependence of contrast sensitivity on field size. At low spatial frequencies, contrast sensitivity decreases with decreasing field size. At high spatial frequencies, the effect of field size gradually disappears. Furthermore, the position of the maximum of the contrast sensitivity function shifts to higher spatial frequencies at smaller field sizes. As the measurements extend to very large angular field sizes, they were used to determine the value of X_{max} used in the model.

3.9.12 Measurements by Rovamo et al. (1992)

Rovamo et al. (1962) measured the contrast sensitivity function with and without two-dimensional static noise. The test object was a square-shaped vertically oriented sinusoidal grating pattern generated on the screen of a high resolution color CRT of which only the green phosphor was used. The luminance was 11 cd/m^2. The size of the test object was constant. It contained 16 cycles with a spatial frequency of 1.5 cycle/cm. The angular spatial frequency was varied by changing the viewing distance. This means that simultaneously the angular field size was varied. The test object had an equiluminous surrounding area of 33 cm × 24 cm. Noise was produced by adding to each pixel a random luminance value from an even distribution with zero mean. The pixel size was 0.53 mm × 0.53 mm on the screen. The relative sigma of the noise was 0.289. The observer looked at the test object with both eyes and without an artificial pupil. The modulation threshold was determined with a 2AFC method where the threshold corresponded with 84% correct response. The contrast sensitivity was not expressed in the usual way as the inverse of the modulation but as the inverse of RMS contrast. This means that the given contrast sensitivity values had to be divided by a factor $\sqrt{2}$. Two experienced subjects (RF and JR), 24 and 37 years of age, served as observers. Only the data from subject JR, the first author, will be used here. This is the same subject as took part in the investigation by Virsu and Rovamo that was treated earlier. At the time of this investigation his age was 25.

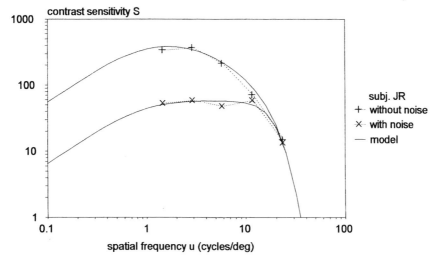

Figure 3.20: Contrast sensitivity function measured by Rovamo et al. (1992) with and without two-dimensional static noise for a square test field containing 16 cycles. Luminance 11 cd/m^2. Binocular viewing with a natural pupil. The solid curves have been calculated with Eq. (3.26). For the calculation of the effect of noise Eqs. (2.42), (2.43) and (2.50) have been used.

Measurements and calculations are shown in Fig. 3.20. The values of σ_0, η, and k used for the calculation were 0.47 arc min, 3.0%, and 3.5, respectively, after correcting k for the difference between 84% and 75% correct response with the aid of Eq. (2.14). For the calculation of the effect of noise, Eqs. (2.42), (2.43) and (2.50) have been used. The agreement between measurements and calculations is very good. The simultaneous fit with the data with and without noise shows that in the model the right value for Φ_0 has been chosen. This value cannot accurately be confirmed by measurements without noise.

3.9.13 Measurements by Rovamo et al. (1993a)

In another investigation, Rovamo et al. (1993a) measured the contrast sensitivity with and without two-dimensional static noise over a wide range of retinal illuminance levels extending over nearly five decades. Only a single spatial frequency was used. The test object was a vertically oriented sinusoidal grating pattern generated on the screen of a high resolution color CRT used in the white mode. The test pattern had a circular shape with a diameter of 20 cm. It contained 20 cycles and was viewed at a distance of 1.15 m. This corresponded with a circular angular field size of 5° and a spatial frequency of 4 cycles/deg. Two-dimensional spatial noise was produced by adding a random luminance value to each pixel from an even distribution with zero mean. The pixel size was 0.42 mm × 0.42 mm on the screen. In one part of the experiment with retinal illumination levels up to 2500 Td, the relative sigma of the noise was 0.4, and in a second part of the experiment with higher retinal illuminance levels, it decreased from this value inversely proportionally with retinal illuminance. The retinal illuminance was varied by using neutral density filters for the lower levels and by adding external light on the screen for the higher levels. Viewing was monocular with the dominant eye of which the pupil was diluted with a drug to a diameter of 8 mm. The modulation threshold was determined with a 2AFC method where the threshold corresponded with 84% correct response. As in the previous investigation, the contrast sensitivity was defined as the inverse of RMS contrast, so that the given contrast sensitivity values had to be divided by a factor $\sqrt{2}$. Two experienced subjects (KT and HK), 25 and 27 years of age, served as observers. The data from subject HK, the second author, will be used here.

Measurements and calculations are shown in Fig. 3.21. In this figure the measurement results of the two parts of the experiment are combined. The sudden increase of the contrast sensitivity above 2500 Td for the measurements with noise is caused by the decrease of noise in the area above this level proportionally with retinal illuminance. The values of σ_0, η, and k used for the calculation were 0.50 arc min, 1.8%, and 3, respectively, after correcting k for the difference between 84% and 75% correct response with the aid of Eq. (2.14). The agreement between measurements and calculations is very good. As for the previous experiment, a simultaneous

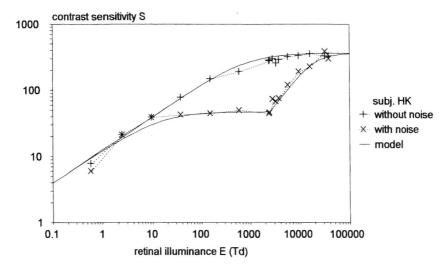

Figure 3.21: Contrast sensitivity with and without two-dimensional static noise measured by Rovamo et al. (1993a) as a function of retinal illuminance. Circular test field with a diameter of 5° and a spatial frequency of 4 cycles/deg. Monocular viewing with a pupil size of 8 mm. In the measurements with noise, the amount of noise was decreased proportionally with the retinal illuminance above 2500 Td. This explains the sudden increase of the contrast sensitivity above this level. The solid curves have been calculated with Eqs. (3.26). For the calculation of the effect of noise Eqs. (2.42), (2.43) and (2.50) have been used.

fit was made for the curves with and without noise. These measurements also confirm the value of Φ_0 used in the model.

The measurements and calculations for the situation without noise clearly show that the contrast sensitivity increases with the square root of retinal illuminance at low illuminance levels, according to the de Vries-Rose law, and that the contrast sensitivity is constant at high illuminance levels, according to Weber's law. The transition takes place at a level of about 1000 Td. This corresponds with a luminance of about 100 cd/m^2 for viewing with a natural pupil. The deviation between measurements and calculations at the lowest illuminance level is caused by the fact that vision at this level is scotopic.

3.9.14 Measurements by Rovamo et al. (1993b)

Rovamo et al. (1993b) also measured the dependence of contrast sensitivity on field size, similar to the measurements made by Carlson. Their measurements extended over a large range of field sizes up to 32°. The luminance was 50 cd/m^2. The test object was a square-shaped vertically oriented sinusoidal grating pattern generated on

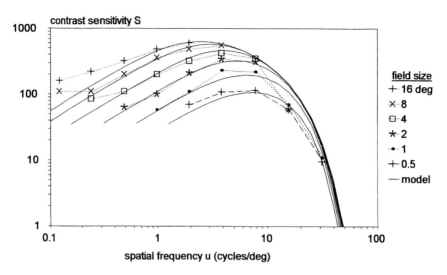

Figure 3.22: Contrast sensitivity function measured by Rovamo et al. (1993b) at a luminance of 50 cd/m² for square test fields with a large range of angular sizes. Binocular viewing with a natural pupil. The solid curves have been calculated with Eq. (3.26).

the screen of a high resolution color CRT with P22 phosphor used in the white mode (color coordinates 0.31, 0.34). The size of the test object varied from 0.5 cm × 0.5 cm to 16 cm × 16 cm and the angular size was further additionally varied by changing the viewing distance. The test object had an equiluminous surrounding area of 27 cm × 20 cm, whereas the further surrounding was completely dark. The observer looked at the test object with both eyes and without an artificial pupil. The pupil size increased with viewing distance from 3.5 mm to 6 mm. For the calculation an average pupil size of 4.9 mm was assumed corresponding with an average illuminance level of 940 Td. The modulation threshold was determined with a 2AFC method where the threshold corresponded with 84% correct response. Six experienced subjects from 24 to 33 years of age took part in the investigation.

Measurements and calculations for one subject, the second author, are shown in Fig. 3.22. The values of σ_0, η, and k used for the calculation were 0.5 arc min, 3%, and 2.7, respectively, after correcting k for the difference between 84% and 75% correct response with the aid of Eq. (2.14). The agreement between measurements and calculations is good, except for the measurements at the lowest spatial frequencies where the data deviate from a linear decay with spatial frequency. Such deviations are not found in the measurements by Carlson shown in Fig. 3.19.

3.9.15 Survey of the measurements

In the model, nine constants play a role, of which six were kept fixed and three were adapted to the measurements, as they could differ for different subjects. For the fixed constants, the values mentioned in section 3.8 were used. A survey of the adapted constants is given in Table 3.1. For measurements with a series of data the same constants have been used for all data. The value of η is influenced by the measurement accuracy of the luminance or the retinal illuminance. In the past the equipment for measuring the luminance was often not very well calibrated. The value of η for the measurements by van Nes & Bouman would have been 3%, instead of 30%, if the measured retinal illuminance was a factor 10 higher.

Table 3.1 σ_0, η, and k values used for the evaluation of the measurements

author	σ_0 (arc min)	η (%)	k
DePalma & Lowry (1962)	0.45	0.5	3.0
Patel (1966)	0.50	3.0	3.1
Robson (1966)	0.53	2.0	4.5
van Nes & Bouman (1967)	0.45 0.54 0.50	30 30 30	2.7 2.7 2.7
Campbell & Robson (1968)	0.53 0.69	2.5 1.0	3.9 3.9
Watanabe et al. (1968)	0.48	3.0	4.0
Sachs et al. (1971)	0.59	3.0	3.7
van Meeteren & Vos (1972)	0.50	3.0	4.0
Howell &Hess (1978)	0.49	3.0	5.0
Virsu & Rovamo (1979)	0.46	2.5	3.2
Carlson (1982)	1.10	3.0	3.8
Rovamo et al. (1992)	0.47	3.0	3.5
Rovamo et al. (1993a)	0.50	1.8	3.0
Rovamo et al. (1993b)	0.50	3.0	2.7

Apart from some exceptions the values of the constants do not much differ from the typical values given in section 3.8.

3.10 Summary and conclusions

In this chapter a model has been given for the spatial contrast sensitivity of the human eye. This model is based on the assumption that the contrast sensitivity is limited by internal noise in the visual system. For this model the basic expressions given in the previous chapter have been used for the evaluation of the effect of noise on the modulation threshold. Furthermore, additional assumptions have been made for the optical MTF of the eye and lateral inhibition. This model gives not only a qualitative description but also a quantitative description of the contrast sensitivity function and its dependence on luminance and field size. The model was compared with a large range of published measurement data and appeared to be in good agreement with them. Measurements with and without external noise gave especially strong support to the model.

In the model, the contrast sensitivity of the eye is explained by effects that mainly take place on the retinal level. This does not mean that there would not exist selective spatial frequency channels for different spatial frequency areas and different orientations, as is often assumed. See, for instance, Sachs et al., 1971. However, from the good agreement between measurements and calculations obtained with the model, it appeared that many aspects of contrast sensitivity can already be explained without the assumption of such channels.

Appendix A. Photon conversion factor

The photon conversion factor p is defined by the number of photons per unit of time, per unit of area, and per unit of luminous flux per angular area entering the eye. It can be derived from basic photometric and physical quantities. See, for instance, Scheibner & Baumgardt (1967). For the luminous flux per angular area use will be made of the Troland that is a measure for the retinal illuminance.

For the energy of a photon holds

$$\varepsilon = h\nu = h\,c/\lambda = 1.9858 \times 10^{-16}/\lambda \ \text{Joule} \tag{3.27}$$

where the wavelength λ is expressed in nm. Taking into account that 1 Watt = 1 Joule/sec, this means that for monochromatic light

$$1\,\text{Watt} = \frac{1}{1.9858} \times 10^{16}\,\lambda \;\text{photons/sec} \qquad (3.28)$$

whereas for non-monochromatic light

$$1\,\text{Watt} = \frac{1}{1.9858} \times 10^{16}\,\frac{\int P(\lambda)\,\lambda\,d\lambda}{\int P(\lambda)\,d\lambda} \;\text{photons/sec} \qquad (3.29)$$

where $P(\lambda)$ is the spectral energy distribution function of the light source.

For monochromatic light and photopic vision holds

$$1\,\text{Watt} = 683\,V(\lambda)\;\text{lumen} \qquad (3.30)$$

where $V(\lambda)$ is the standard spectral sensitivity distribution for photopic vision adopted by the CIE (Commission International de l'Éclairage) in 1924. This function has as maximum value 1 at 555 nm. For non-monochromatic light

$$1\,\text{Watt} = 683\,\frac{\int P(\lambda)\,V(\lambda)\,d\lambda}{\int P(\lambda)\,d\lambda}\;\text{lumen} \qquad (3.31)$$

Combining these expressions with Eqs. (3.28) and (3.29), respectively, gives for monochromatic light

$$1\,\text{lumen} = 7.373 \times 10^{12}\,\lambda/V(\lambda)\;\text{photons/sec} \qquad (3.32)$$

and for non-monochromatic light

$$1\,\text{lumen} = 7.373 \times 10^{12}\,\frac{\int P(\lambda)\,\lambda\,d\lambda}{\int P(\lambda)\,V(\lambda)\,d\lambda}\;\text{photons/sec} \qquad (3.33)$$

From the international definition of the Troland given by Eq. (3.15) follows

$$1\,\text{Troland} = 1\,\text{cd/m}^2 \times 1\,\text{mm}^2 = 10^{-6}\,\text{cd} = 10^{-6}\,\text{lumen/sterad} \qquad (3.34)$$

To express the photon conversion factor in the same units of the angular object size that are used in the model, the sterad has to be replaced by the angular area in degrees. This gives

$$1\,\text{Troland} = \frac{10^{-6}}{\left(\dfrac{180}{\pi}\right)^2}\;\text{lumen/deg}^2 = 3.0462 \times 10^{-10}\;\text{lumen/deg}^2 \qquad (3.35)$$

For monochromatic light, one obtains by combining this equation with Eq. (3.32)

$$1\,\text{Troland} = 2.246 \times 10^3\,\lambda/V(\lambda)\;\text{photons/sec/deg}^2 \qquad (3.36)$$

and for non-monochromatic light, one obtains by combining this equation with Eq. (3.33)

$$1 \, \text{Troland} = 2.246 \times 10^3 \, \frac{\int P(\lambda) \lambda \, d\lambda}{\int P(\lambda) V(\lambda) d\lambda} \quad \text{photons/sec/deg}^2 \qquad (3.37)$$

Photons at the extreme ends of the visual spectrum have only a small contribution to the viewing process. Multiplying the contribution of each part of the spectrum by $V(\lambda)$ is, therefore, more convenient. In this way, all contributions are weighted by their relative sensitivity with respect to the most sensitive part of the spectrum where $V(\lambda) = 1$. For this *photopic weighted* number of photons holds for monochromatic light

$$1 \, \text{Troland} = 2.246 \times 10^3 \, \lambda \, \text{photons/sec/deg}^2 \qquad (3.38)$$

and for non-monochromatic light

$$1 \, \text{Troland} = 2.246 \times 10^3 \, \frac{\int P(\lambda) V(\lambda) \lambda \, d\lambda}{\int P(\lambda) V(\lambda) d\lambda} \quad \text{photons/sec/deg}^2 \qquad (3.39)$$

For photopic viewing conditions, the photopic weighted number of photons has to be used for the photon conversion factor. This gives for the photon conversion factor for monochromatic light

$$p = 2.246 \times 10^3 \, \lambda \, \text{photons/sec/deg}^2/\text{Td} \qquad (3.40)$$

and for the photon conversion factor for non-monochromatic light

$$p = 2.246 \times 10^3 \, \frac{\int P(\lambda) V(\lambda) \lambda \, d\lambda}{\int P(\lambda) V(\lambda) d\lambda} \quad \text{photons/sec/deg}^2/\text{Td} \qquad (3.41)$$

For scotopic viewing conditions a *scotopic weighted* photon conversion factor has to be used. Then the $V(\lambda)$ function in the numerator of Eq. (3.41) has to be replaced by the $V'(\lambda)$ function for scotopic vision standardized by the CIE in 1951. This function has as a maximum value 1 at 507 nm. The right-hand side of Eq. (3.40) for monochromatic light has simultaneously to be multiplied by $V'(\lambda)/V(\lambda)$.

Usually the luminance is measured in photopic units based on the $V(\lambda)$ function. However, if the luminance is measured in scotopic units based on the $V'(\lambda)$ function, Eq. (3.30) has to be replaced by

$$1 \, \text{Watt} = 1700 \, V'(\lambda) \, \text{lumen} \qquad (3.42)$$

where also the numerical constant is different because of the different shape of the $V'(\lambda)$ function. Consequently, the factor 2.246×10^3 in Eqs. (3.40) and (3.41) has to be replaced by 0.9024×10^3 and $V(\lambda)$ in the denominator of Eq. (3.41) has to be replaced by $V'(\lambda)$, whereas the right-hand side of Eq. (3.40) has to be multiplied by $V(\lambda)/V'(\lambda)$.

Note that in addition to the standard $V(\lambda)$ curve, which was adopted in 1924 and is in principle only valid for fields with a diameter of 2°, the CIE adopted in 1963 a $V_{10}(\lambda)$ curve to be used for larger fields with a diameter of 10°. The reason for the difference is the presence of macular pigment in the fovea, which mainly absorbs blue light. Therefore, the main difference between the curves is the sensitivity for blue. Outside the fovea the sensitivity for blue light is about a factor two higher. Normal light-meters are based on the standard $V(\lambda)$ curve for 2°.

A survey of the photon conversion factor for different light sources is given in Table 3.2. They have been used for the evaluation of the contrast sensitivity measurements that were compared with the given model. The mentioned values for scotopic viewing are given as general information, as the model can only be used for photopic viewing conditions.

Table 3.2 Photon conversion factor p for different light sources
in 10^6 photons/sec/deg^2/Td

light source	photopic vision		scotopic vision	
	per phot. Td	per scot. Td	per phot. Td	per scot. Td
monochromatic blue 450 nm	1.011	0.0330	12.42	0.406
monochromatic green 525 nm	1.179	0.4204	1.329	0.474
monochromatic green 555 nm	1.247	1.229	0.508	0.501
monochromatic red 650 nm	1.460	87.34	0.0099	0.586
illumin. A (color temp. 2854 K)	1.285	0.891	0.671	0.466
P1 (green CRT phosphor)	1.201	0.467	1.212	0.472
P31 (blue-green CRT phosphor)	1.221	0.510	1.116	0.466
P4 (white CRT phosphor)	1.240	0.432	1.287	0.449

References

Barten, P.G.J. (1992). Physical model for the contrast sensitivity of the human eye. *Human Vision, Visual Processing, and Digital Display III, Proc. SPIE*, **1666**, 57-72.

Blommaert, F.J.J., Heijnen, H.G.M., and Roufs, J.A.J. (1987). Point spread functions and detail detection. *Spatial Vision*, **2**, 99-115.

Bouma, H. (1965). Receptive systems. Mediating certain light reactions of the pupil of the human eye. Ph.D. Thesis, Technical University Eindhoven, The Netherlands. *Philips Research Reports Supplements*, **5**.

Campbell, F.W. & Green, D.G. (1965) Monocular versus binocular visual acuity. *Nature*, **208**, No. 5006, 191-192.

Campbell, F.W. & Robson, J.G. (1968). Application of Fourier analysis to the visibility of gratings. *Journal of Physiology*, **197**, 551-566.

Carlson, C.R. (1982). Sine-wave threshold contrast-sensitivity function: dependence on display size. *RCA Review*, **43**, 675-683.

Crawford, B.H. (1936). The dependence of pupil size upon external light stimulus under static and variable conditions. *Proceedings Royal Society*, **B 121**, 376-395.

de Groot, S.G. & Gebhard, J.W. (1952). Pupil size determined by adapting luminance. *Journal of the Optical Society of America*, **42**, 492-495.

DePalma, J.J. & Lowry, E.M. (1962). Sine-wave response of the visual system. II. Sine-wave and square-wave contrast sensitivity. *Journal of the Optical Society of America*, **52**, 328-335.

de Vries, H. (1943). The quantum character of light and its bearing upon threshold of vision, the differential sensitivity and visual acuity of the eye. *Physica*, **10**, 553-564.

Enroth-Cugell, C. & Robson, J.G. (1966). The contrast sensitivity of retinal ganglion cells of the cat. *Journal of Physiology*, **187**, 517-522.

Field, D.J. (1987). Relations between the statistics of natural images and the response properties of cortical cells. *Journal of the Optical Society of America A*, **4**, 2379-2394.

Howell, E.R. & Hess, R.F. (1978). The functional area for summation to threshold for sinusoidal gratings. *Vision Research*, **18**, 369-374.

Jacobs, D.H. (1944). The Stiles-Crawford effect and the design of telescopes. *Journal of the Optical Society of America*, **34**, 694.

Kumnick, L.S. (1954). Pupillary psychosensory restitution and aging. *Journal of the Optical Society of America*, **44**, 735-741.

le Grand, Y. (1969). Light, colour and vision. 2nd edition, Chapman and Hall, London.

Moon, P. & Spencer, D.E. (1944). On the Stiles-Crawford effect. *Journal of the Optical Society of America*, **34**, 319-329.

Papoulis, A. (1968). Systems and transforms with applications in optics. McGraw-Hill, New York-St. Louis-San Francisco-Toronto-London-Sydney.

Patel, A.S. (1966). Spatial resolution by the human visual system. The effect of mean retinal illuminance. *Journal of the Optical Society of America*, 56, 689-694.

Robson, J.G. (1966). Spatial and temporal contrast-sensitivity functions of the visual system. *Journal of the Optical Society of America*, 56, 11417-1142.

Rose, A. (1942). The relative sensitivities of television pickup tubes, photographic film, and the human eye. *Proceedings of the IRE*, 30, 293-300.

Rose, A (1948). The sensitivity performance of the human eye on an absolute scale. *Journal of the Optical Society of America*, 38, 196-208.

Rovamo, J., Franssila, R., and Näsänen, R. (1992). Contrast sensitivity as a function of spatial frequency, viewing distance and eccentricity with and without spatial noise. *Vision Research*, 32, 631-637.

Rovamo, J., Kukkonen, H., Tiippana, K., and Näsänen, R. (1993a). Effects of luminance and exposure time on contrast sensitivity in spatial noise. *Vision Research*, 33, 1123-1129.

Rovamo, J., Luntinen, O., and Näsänen, R. (1993b). Modelling the dependence of contrast sensitivity on grating area and spatial frequency. *Vision Research*, 33, 2773-2788.

Sachs, M.B., Nachmias, J., Robson, J.G. (1971). Spatial-frequency channels in human vision. *Journal of the Optical Society of America*, 61, 1176-1186.

Schade, O. (1956). Optical and photoelectric analog of the eye. *Journal of the Optical Society of America*, 46, 721-739.

Scheibner, H. & Baumgardt, E. (1967). Sur l'emploi en optique physiologique des grandeurs scotopiques. *Vision Research*, 7, 59-63.

Stiles, W.S. & Crawford, B.H. (1933). The luminance efficiency of rays entering the eye at different points. *Proceedings Royal Society*, B 112, 428-450.

van Meeteren, A. (1973). Visual aspects of image intensification. Ph.D. Thesis, Utrecht University, Utrecht, The Netherlands.

van Meeteren, A. (1978). On the detective quantum efficiency of the human eye. *Vision Research*, 18, 257-267.

van Meeteren, A. & Vos, J.J. (1972). Resolution and contrast sensitivity at low luminance levels. *Vision Research*, 12, 825-833.

van Nes, F.L. (1968). Experimental studies in spatiotemporal contrast transfer by the human eye. Ph.D. Thesis, Utrecht University, Utrecht, The Netherlands.

van Nes, F.L. & Bouman, M.A. (1967). Spatial modulation transfer in the human eye. *Journal of the Optical Society of America*, 57, 401-406.

Virsu, V. & Rovamo, J. (1979). Visual resolution, contrast sensitivity, and the cortical magnification factor. *Experimental Brain Research*, 37, 475-494.

Vos, J.J., Walraven, J., and van Meeteren, A. (1976) Light profiles of the foveal image of a point source. *Vision Research*, 16, 215-219.

Watanabe, A., Mori, T., Nagata S., and Hiwatashi, K. (1968) Spatial sine-wave responses of the human visual system. *Vision Research*, **9**, 1245-1263.

Young, R.A. (1991). Oh say, can you see? The physiology of vision. *Human Vision, Visual Processing, and Digital Display II, Proc. SPIE*, **1453**, 92-123.

Chapter 4

Extension of the contrast sensitivity model to extra-foveal vision

4.1 Introduction

The spatial contrast sensitivity model given in the previous chapter is restricted to the normal situation of foveal vision. At foveal vision, the eyes of an observer are directed to an object in such a way that the center of the object is imaged on the center of the retina where the contrast sensitivity of the eye is maximum. This process is called *foveal fixation*. In daily practice, for instance in traffic, it is also important that the eye can observe objects that are outside the area on which the main attention is concentrated. In this chapter the model given in the previous chapter will be extended, so that it can also be used for extra-foveal vision. Outside the fovea, the contrast sensitivity and the resolution of the eye is much less. To measure the local contrast sensitivity outside the fovea, a marker is usually placed in the object plane and the observer is asked to fixate his eye on this marker, while the actual object is placed at some distance from the marker. This distance is usually expressed in an angular measure called *eccentricity* and the contrast sensitivity is measured as a function of eccentricity. As the instruction to fixate the eye on the marker is not always easy to follow, extra-foveal contrast sensitivity measurements usually show more spread than foveal contrast sensitivity measurements. The foveal area has a diameter of about 1°. Although the center of the object at foveal vision is imaged on the fovea, a large part of the image will usually also cover the retinal area outside the fovea. At extra-foveal vision, the center of the object is imaged outside the fovea, but a part of the image can still cover the foveal area.

For the extension of the contrast sensitivity model to extra-foveal vision, it is sufficient to adapt the constants used in the model as a function of eccentricity. It may be assumed that the variation of the constants with eccentricity is caused by the density variation of the cones and ganglion cells over the retina. Therefore, first some approximation formulae will be given for the density distribution of these types of cells over the retina, then the effect of these cell types on the contrast sensitivity will be analyzed, and finally the so extended contrast sensitivity model will be compared with various published measurements.

4.2 Density distribution of retinal cells

For photopic vision the density variation of cones and ganglion cells over the retina plays the most important role at the variation of contrast sensitivity with eccentricity. However, the density distribution of the rods will also be treated in this section to obtain a good insight into the composition of the retina. Before going into detail about the density distribution of the different cell types, first some geometrical relations will be given that can be used for all types of retinal cells.

4.2.1 Geometrical relations

From microscopical investigations of the retina, it appears that the cells are largely arranged in hexagonal patterns with randomly different orientations. See, for instance, Polyak (1957, pp. 268-271) and Curio et al. (1987, Fig. 1). In principle a strictly hexagonal array is not possible, as the cell density varies over the retina. It can, therefore, be found only in local areas where the density is approximately constant. However, the hexagonal pattern can still be used as description of the average local situation.

Under this assumption, the distance s between two neighbouring rows of cells is given by

$$s = \frac{1}{2}\sqrt{3}\, d \qquad\qquad (4.1)$$

where d is the center-to-center distance of the cells. See Fig. 4.1. The available surface area A per cell is

$$A = s d = \frac{2}{\sqrt{3}}\, s^2 \qquad\qquad (4.2)$$

and the density N of the cells per unit area is

$$N = \frac{1}{A} = \frac{\frac{1}{2}\sqrt{3}}{s^2} \qquad\qquad (4.3)$$

The row spacing s is usually expressed in arc min of the corresponding visual angle in the object space, and the density N is usually expressed in the number of cells per $\deg^2$ of visual angle. In this case

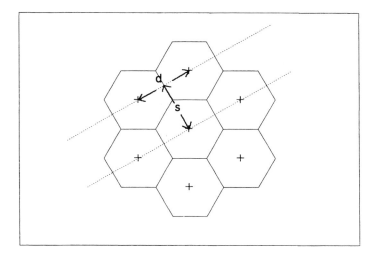

Figure 4.1: Schematic view of the hexagonal structure of the cells on the retina. d is te center-to-center distance of the cells and s is the row spacing of the cells, the distance between neighbouring rows of cells. Their mutual relation is given by Eq. (4.1).

$$N = \frac{\frac{1}{2}\sqrt{3} \times 60^2}{s^2} = \frac{3118}{s^2} \quad \text{deg}^{-2} \tag{4.4}$$

In biological publications the cell density is usually expressed in cells per square mm on the surface of the retina. The conversion factor from mm on the retina to the visual angle in the object space amounts to about 0.291 mm/deg for the adult human eye. See Williams (1988, footnote, page 441). So, for the human eye holds

$$N = \frac{\frac{1}{2}\sqrt{3} \times 60^2}{0.291^2} \frac{1}{s^2} = \frac{36817}{s^2} \quad \text{mm}^{-2} \tag{4.5}$$

where s is still expressed in arc min. The cell density N varies with retinal eccentricity. This eccentricity is usually expressed in degrees.

4.2.2 Cone density distribution

Measurements of the density of cones and rods as a function of retinal eccentricity have already been made in the first part of the 19th century by Østerberg (1935). He investigated tissues of the human retina with the aid of a microscope. His measurements of the cone density have more recently been confirmed by measurements of Williams and co-workers (Coletta & Williams, 1987, Williams, 1988) who measured this density in the living eye. They made these measurements at various eccentricities by using the interference between rows of cones and a grating pattern that was projected on the retina. This interference was visible as a moire pattern. Observers varied the spatial frequency of the grating pattern until the moire pattern was as coarse as possible. This method could be used for eccentricities up to 15°. At larger eccentricities the regularity of the hexagonal array is too much disturbed by local variations to obtain reliable results.

Fig. 4.2 shows the cone density plotted as a function of eccentricity on a double logarithmic scale. The measured data are from Østerberg, Coletta & Williams, and Williams. The measurements by Østerberg were given in cells per mm² at eccentricities in mm. For the conversion from mm to degrees the factor of 0.291 mm/deg mentioned in the section 4.2.1 was used. The measurements by Coletta & Williams and Williams were expressed in row spacing. The cone density was calculated from the row spacing with Eq. (4.4). The measurements by Coletta and Williams are the individual results of three observers and the measurements by Williams are the average results of eight observers. From the measurements by Coletta the data for eccentricities above 16° have been omitted for the reasons mentioned above. The figure shows that the different types of measurements agree very well with each other. The curve through the data points in the figure was calculated with the following empirical approximation formula:

$$N_c = N_{c0} \left(\frac{0.85}{1 + (e/0.45)^2} + \frac{0.124}{1 + (e/6)^2} + 0.026 \right) \qquad (4.6)$$

where N_c is the cone density, N_{c0} is the cone density in the center of the retina and

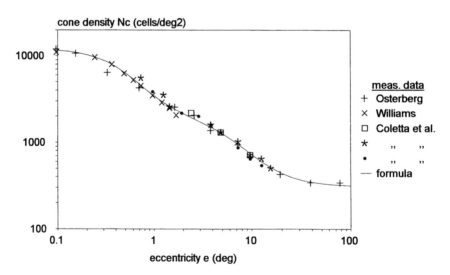

Figure 4.2: Cone density as a function of retinal eccentricity derived from measurements by Østerberg (1935), Williams (1988), and Coletta & Williams (1987). The curve through the data represents the approximation formula given by Eq. (4.6).

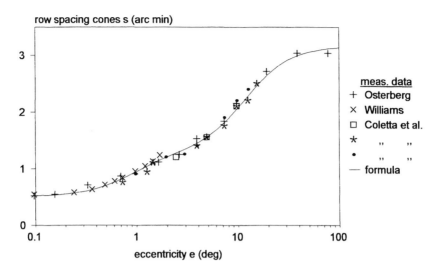

Figure 4.3: Row spacing of cones as a function of retinal eccentricity for the measurement data of Fig. 4.2. The curve through the data points has been calculated with Eqs. (4.6) and (4.4).

e is the eccentricity in degrees. In this formula, the cell density is described as the sum of three overlapping distributions: the density variation in the fovea, the density variation in the area between fovea and periphery, and the density in the periphery. These densities are indicated in this order by the three terms between the brackets. The formula appears to describe the data points very well. The value used for N_{c0} was 12,000 cells/deg^2. From the equations given in section 4.2.1 follows that this corresponds with a density of 142,000 cells/mm^2, a row spacing *s* of 0.51 arc min, and a distance *d* between the center of the cones of 0.59 arc min for the center of the retina. Fig. 4.3 shows the row spacing of the cones calculated with Eq. (4.4) for the data given in Fig. 4.2 as a function of the eccentricity. From this figure, it can be seen that the cone spacing varies from about 0.5 arc min in the fovea to about 3.2 arc min in the periphery.

4.2.3 Rod density distribution

Although the rods have no direct influence on the contrast sensitivity at the photopic levels to which we restrict us, a short discussion of their density distribution will be given here as general information about the composition of the retina. At this moment, the best available data of the density distribution of the rods are measurements by Østerberg (1935) that were made together with the measurements of the cones mentioned in the previous section. These measurements are shown in Fig. 4.4. The curve through the data points was calculated with the following empirical

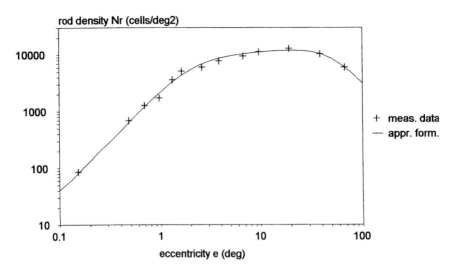

Figure 4.4: Rod density as a function of retinal eccentricity measured by Østerberg (1935). The curve through the data points has been calculated with an empirical approximation formula given by Eq. (4.7).

approximation formula:

$$N_r = 12{,}000 \left(1 - \frac{0.85}{1 + (e/2.0)^2} - \frac{0.15}{0.15 + 0.85/(1 - e/20)^2} \right) \quad \text{cells/deg}^2 \quad (4.7)$$

where N_r is the rod density, and e is the eccentricity in degrees. From a comparison with Fig. 4.2, it appears that the rod density varies over the retina almost complementary with the cone density. The rod density increases from about zero in the center of the retina to a maximum of about 12,000 cells/deg^2 at an eccentricity of 20°, which is equal to the cone density in the center of the retina. At this eccentricity the rod spacing is also about equal to the cone spacing in the center of the retina.

4.2.4 Ganglion cell density distribution

The ganglion cells in the retina are the cells from which the visual information is transported to the brain. They consist of different types that have different tasks. They can be distinguished (See Henry & Vidyasagar, 1991) in *P-cells*, *M-cells* and *K-cells*, of which the P-cells are connected with the *parvocellular layer* of the *lateral geniculate nucleus* (LGN) in the brain, the M-cells are connected with the *magnocellular layer* of the LGN and the K-cells are connected with the *koniocellular group* in the LGN. From the LGN the information is further transported to the *visual cortex*, the part of the brain where the visual information is received. According to Lee (1996), for the primate retina, 80% of the ganglion cells are P-cells, 10% are M-cells and the remaining 10% are K-cells. The P-cells are responsible for the color information, the M-cells are responsible for the luminance information and the function of the K-cells is still unknown. The M-cells can be distinguished in *on-center M-cells* and *off-center M-cells*, which names say that the center of the receptive field is activated by "light on" and "light off," respectively. Each of them forms 5% of the total amount of ganglion cells. It is assumed here that this also holds for the human retina. The on-center M-cells convey the luminance information and are, therefore, responsible for the contrast sensitivity of the eye. They pool the information of several photo-receptors. Therefore, not the density of the photo-receptors itself, but the density of this type of ganglion cells determines the resolution of the eye.

According to Wässle et al. (1990), who investigated the retina of macaque monkeys, there are about three ganglion cells per cone in the center of the retina. As about 5% of these cells are on-center M-cells, there are about 0.15 on-center M-cells per cone, or about six cones per on-center M-cell. Because of the resemblance between the eye of the macaque monkey and the human eye, it may be assumed that this number is also valid for the human eye. Using the cone density of 12,000 cells/deg^2 in the center of the retina mentioned in the previous section, the density of the ganglion cells is approximately $3 \times 12,000 = 36,000$ cells/deg^2 in the center of the retina and the density of the on-center M ganglion cells is approximately $0.05 \times 36,000 = 1,800$ cells/deg^2 in the center of the retina.

Outside the fovea the ganglion cell density decreases more steeply than the cone density. The decrease is different for the four main directions that are usually distinguished: the *nasal direction* and the *temporal direction* along the horizontal meridian of the retina, and the *superior direction* and the *inferior direction* along the vertical meridian of the retina. These are the directions of nose and temple, and the upward and downward direction, respectively. Because the image on the retina is an inversed image of the observed object, these directions correspond with opposite directions in the observed field. Curcio & Allen (1990) made anatomical measurements of the density distribution of ganglion cells in human eyes along these directions. They used eyes of eye bank donors, which they measured within three hours after the death of the donor. The donors were less than 37 years of age. Fig. 4.5 shows the results for the average of six eyes. The given densities are the densities of the total number of ganglion cells. To obtain the densities of the on-center M-cells, 5% of this density has to be taken, assuming that this percentage is also valid for the human eye and is constant over the retina. Data for an eccentricity smaller than four degrees have been omitted in the figure, because the ganglion cells at these eccentricities are displaced with respect to the cones with which they are connected. This displacement, which is usually called *Henle effect*, is caused by an overcrowding of cells in the fovea. There are also no measurements in the nasal area at an eccentricity of 14° because of the presence of a hole in the retina called the *blind spot*, which serves for the passage of the nerve fibers. In the remaining part of the nasal area, the

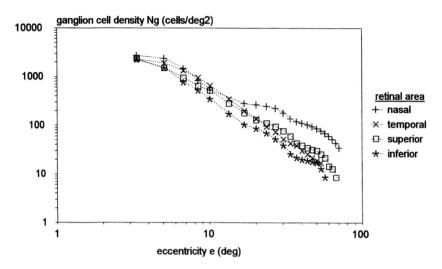

Figure 4.5: Ganglion cell density as a function of eccentricity measured by Curcio & Allen (1990) for different areas of the human retina.

ganglion cell density is clearly higher than in the other areas. This can be explained by the need during the evolution of the eye to have good vision in sideways directions. For probably the same reasons, the density in the inferior area of the retina is less than in the other areas, because the visual field of this area corresponds with the skies, for which less resolution is needed.

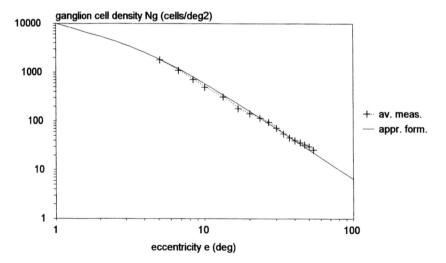

Figure 4.6: Ganglion cell density as a function of eccentricity. Average of the measurements for the different areas given in Fig. 4.5. The solid curve represents the approximation formula given by Eq. (4.8) with $e_g = 3.3°$.

The average density variation in the different areas of the retina can be described by an approximation formula that has partly the similar form as Eq. (4.6) and is given by

$$N_g = N_{g0} \left(\frac{0.85}{1 + (e/0.45)^2} + \frac{0.15}{1 + (e/e_g)^2} \right) \qquad (4.8)$$

where N_g is the ganglion cell density, N_{g0} is the ganglion cell density in the center of the retina, e is the eccentricity in degrees, and e_g is a constant that can be different for different subjects. In this expression the first term between the brackets is equal to the first term between the brackets of Eq. (4.6), because it is assumed here that the ratio between ganglion cells and cones is constant in the foveal area. The function is shown in Fig. 4.6 with the average of the measurements for the different areas given in Fig. 4.5. For N_{g0}, the value of 36,000 cells/deg mentioned in the previous section is used and for e_g a value of 3.3°. The figure shows that this approximation formula gives a good description of the average of the measured data.

4.3 Effect of eccentricity on the different constants used in the model

To extend the use of the contrast sensitivity model to extra-foveal vision, assumptions have to be made about the effect of the eccentricity on the constants for the different parameters used in the model. The variation of these parameters may be assumed to be connected with the variation in density of the different retinal elements with eccentricity.

4.3.1 Effect of eccentricity on resolution

As mentioned in section 3.3 of the preceding chapter, the optical MTF used in the model does not include only the optical effect of the eye lens and other optical effects, but also the effect of the discrete structure of the retina. The value of σ in Eq. (3.6), which describes the optical MTF, is the standard deviation of the line-spread function resulting from the convolution of the different elements by which the optical MTF of the eye is determined. To take the effect of the structure of the retina explicitly into account σ_0 in Eq. (3.7) may be written as

$$\sigma_0 = \sqrt{\sigma_{00}^2 + \sigma_{ret}^2} \qquad (4.9)$$

where σ_{ret} is the standard deviation of the line-spread function caused by the discrete structure of the retina and σ_{00} is the standard deviation of the remaining parts of σ_0. In this equation, it is assumed that the standard deviation of the line-spread function resulting from the convolution is equal to the square root of the sum of squares of the standard deviations of the different elements. By inserting Eq. (4.9) in Eq. (3.7), this equation is modified into

$$\sigma = \sqrt{\sigma_{00^2} + \sigma_{ret}^2 + (C_{ab}\, d)^2} \qquad (4.10)$$

σ_{ret} can be derived from the size of an elementary area that delivers information to the brain. As already mentioned in the previous section, the on-center M-cells pool the information of several cones for a further transport to the brain. This means that σ_{ret} is determined by the receptive field of these cells. For the ganglion cells, it may be assumed that these areas are arranged in a hexagonal array similar to that of the cones. The spacing s_g between rows of ganglion cells can then be derived from the density of the ganglion cells. Using Eq. (4.3), one obtains

$$s_g = \sqrt{\frac{\frac{1}{2}\sqrt{3}}{N_g}} \qquad (4.11)$$

Rows of on-center M-cells form the line-spread function that determines the effect of the retinal structure on the optical MTF of the eye. The 50% width of the line-spread function is equal to the row spacing of these cells. This is shown in Fig. 4.7 where it is assumed that the collected information is uniformly distributed over the hexagonal receptive field of the ganglion cells. The line-spread function formed by the rows of ganglion cells is shown by the shaded area at the bottom of this figure. The standard deviation of this function can be calculated from an integration over the width of the

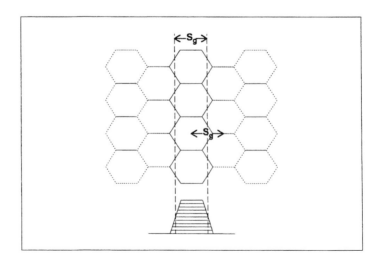

Figure 4.7: Schematic view of a row of on-center M-cells which determines the effect of the retinal structure on the optical MTF of the eye. The line-spread function formed by this row is indicated by the shaded area at the bottom of the figure. The 50% width of this function is equal to the row spacing s_g of these cells.

intensity distribution. From the geometrical configuration given in Fig. 4.7 can in this way be derived that

$$\sigma_{ret} = \frac{s_g}{3\sqrt{6/5}}$$ (4.12)

where s_g is the row spacing of the on-center M-cells, and it is further assumed that the collected information is uniformly distributed over the hexagonal receptive fields of these cells and that an overlap of these fields may be neglected. By replacing s_g in this expression by N_g with the aid of Eq. (4.11), one obtains

$$\sigma_{ret} = \frac{1}{\sqrt{7.2\sqrt{3}\,N_g}}$$ (4.13)

where N_g is now the density of the on-center M-cells. Applying this equation to the center of the retina, where the density of these ganglion cells is approximately 1,800 cells/deg^2, gives σ_{ret} = 0.40 arc min. In section 3.3 of Chapter 3, it was mentioned that σ_0 is approximately 0.5 arc min for observers with good vision. From the value of σ_{ret} follows that for these observers $\sigma_{00} = \sqrt{(0.5^2 - 0.4^2)}$ = 0.30 arc min. This means that for foveal vision and a small size of the eye pupil, the contribution of the retinal structure of the eye on the resolution is about equal to that of other factors. The development of the eye during evolution seems, therefore, to be well balanced. If, instead of the density of the on-center M ganglion cells, the density of the cones was used in the calculation, σ_{ret} would have been 0.16 arc min. If, furthermore, the lens aberrations were neglected, the sharpness of the eye would have been a factor 0.5/0.16 ≈ 3 times better. This illustrates that it is wrong to take only the density of the cones into account for the calculation of the resolution of the eye. The pooling of information by the ganglion cells over different cones forms obviously a stabilizing element in the processing of information.

At increasing distance from the center of the retina, σ_{ret} increases because of the decrease of the ganglion cell density. This means that at some distance from the center, σ_{ret} becomes dominant and the resolution of the eye becomes completely determined by the density of the ganglion cells, as the other terms in Eq. (4.10) remain nearly constant or increase less with eccentricity. With the aid of Eq. (4.13) the value of σ of the optical MTF of the eye can then be calculated from the ganglion cell density at different eccentricities. Inversely, the ganglion cell density at different eccentricities can be calculated from the value of σ derived from measurements of the contrast sensitivity at these eccentricities. This method will be used in section 4.4 for an analysis of extra-foveal contrast sensitivity measurements.

4.3.2 Effect of eccentricity on neural noise

As already mentioned in section 3.5 of the preceding chapter, neural noise is assumed to be caused by statistical fluctuations of the signals in the nerve fibers by which the luminance information is transported to the brain. By inversion of Eq. (3.18) one obtains

$$\Phi_0 = \sigma^2 \Delta x \, \Delta y \, \Delta t = \frac{\sigma^2 \Delta t}{N_g} \qquad (4.14)$$

where $\Delta x \Delta y$ has been replaced by $1/N_g$ and N_g is the density of the on-center M-cells. In this expression σ is the relative standard deviation of the signal transported by an individual nerve fiber. If it is assumed that σ and Δt do not vary with eccentricity, this expression says that the spectral density of neural noise varies inversely proportionally to the density of the ganglion cells. This means that the spectral density of neural noise can be expressed as a function of eccentricity by the following expression:

$$\Phi_0(e) = \Phi_0 \frac{N_{g0}}{N_g} \qquad (4.15)$$

where Φ_0 at the right-hand side of this expression is the value of Φ_0 at foveal vision, given in the previous chapter, and N_g can be calculated with Eq. (4.8), or can be derived from a best fit with extra-foveal contrast sensitivity measurements.

4.3.3 Effect of eccentricity on lateral inhibition

From the equations given in section 3.6 of the previous chapter can be calculated that the lowpass filter used in the lateral inhibition process extends over a receptive field with a radius of about 3 arc min. See Fig. 3.3. This holds for foveal vision. At increasing eccentricity not only the mutual distance between the cones increases, but also the size of the receptive field. The spatial frequency u_0, where the lateral inhibition ceases, varies inversely proportionally with the size of the receptive field and will therefore shift to lower spatial frequencies. One would expect that the diameter of the receptive field would increase proportionally with the distance between the ganglion cells, so that the number of ganglion cells involved in the lateral inhibition process remains about constant. This means that u_0 would vary proportionally with $\sqrt{N_g}$. However, as the ganglion cell density decreases rather steeply with increasing eccentricity, this would lead to a very large field for the lateral inhibition at high eccentricities and to a very small value of u_0. This is not in agreement with measurements. Therefore, it has to be assumed that the receptive field of the inhibition process contains a decreasing number of ganglion cells with increasing eccentricity so that the decrease of u_0 with eccentricity is limited. From an analysis of the extra-foveal contrast sensitivity measurements that will be given in section 4.4 of this chapter, it appears that the variation of u_0 with eccentricity may approximately be described by the following empirical formula:

$$u_0(e) = u_0 \left(\frac{N_g}{N_{go}} \right)^{0.5} \left(\frac{0.85}{1 + (e/4)^2} + \frac{0.13}{1 + (e/20)^2} + 0.02 \right)^{-0.5} \qquad (4.16)$$

where u_0 at the right-hand side of this expression is the value of the constant at foveal vision, given in the previous chapter, and e is the eccentricity in degrees. The decrease of the number of ganglion cells used in the inhibition process is described by the last factor of this equation.

4.3.4 Effect of eccentricity on quantum efficiency

The quantum efficiency of the eye is sometimes defined by the chance that a photon falling on a photo-receptor causes an excitation of this receptor. However, according to the definition used here, the quantum efficiency is defined by the average number of photons causing an excitation of the photo-receptors, divided by the number of photons entering the eye. See section 3.4. This means that the quantum efficiency also depends on the area of the retina covered by the photo-receptors. In the center of the fovea hardly any rods are present, so that nearly the complete area is covered with cones. This means that the quantum efficiency in the center of the fovea is very low for the rods, but is maximum for the cones. Outside the center of the retina the area covered by the rods increases, and the area covered by the cones consequently decreases, so that the quantum efficiency of the rods increases and the quantum efficiency of the cones decreases. In the central area of the retina the quantum efficiency of the cones is flattened by the presence of macular pigment in the foveal area, which acts as a filter for blue light. From an evaluation of contrast sensitivity measurements which will be given in section 4.4, it appears that the dependence of

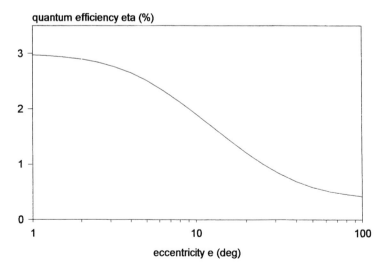

Figure 4.8: Photopic quantum efficiency as a function of retinal eccentricity calculated with Eq. (4.18) and assuming a quantum efficiency of 3% in the center of the retina.

the quantum efficiency of the cones with eccentricity can approximately be described by the following empirical formula:

$$\eta(e) = \eta \left(\frac{0.4}{1 + (e/7)^2} + \frac{0.48}{1 + (e/20)^2} + 0.12 \right) \qquad (4.17)$$

where η at the right-hand side of this expression is the quantum efficiency at foveal vision and e is the eccentricity in degrees. Fig. 4.8 shows a plot of this function for the typical situation that the quantum efficiency of the cones is 3% in the center of the retina.

4.3.5 Effect of eccentricity on the maximum integration area

As was mentioned in section 2.4 of Chapter 2, the area where a signal can be compared with noise, is limited by a maximum angular size X_{max} and by a maximum number of cycles N_{max}. For foveal vision X_{max} is $12°$. For extra-foveal vision, it may be assumed that about a constant number of neurons is involved in the integration process at increasing eccentricity. This does not mean that the maximum size of the integration area increases proportionally with the distance between the ganglion cells at increasing eccentricity. A proportional increase can be true for a small part of the area, but will not be true for the maximum area as a whole. This is caused by the influence of the high ganglion cell density in the foveal area. At foveal vision, the foveal area with its diameter of $1°$ is only a small part of the total maximum field size of $12° \times 12°$. This means that for extra-foveal vision, the maximum size of the integration area increases much slower than proportionally to the ganglion cell distance in the center of the object. In many experiments the field size of the object is much smaller than $12° \times 12°$. For these experiments an increase of X_{max} at larger eccentricities is irrelevant. In only some of the experiments that will be given in section 4.4 of this chapter, a large field size was used. From these experiments, it appeared that the variation of X_{max} with eccentricity can approximately be described by the following expression:

$$X_{max}(e) = Y_{max}(e) = X_{max} \left(\frac{0.85}{1 + (e/4)^2} + \frac{0.15}{1 + (e/12)^2} \right)^{-0.5} \qquad (4.18)$$

where X_{max} at the right-hand side of this expression is the value of X_{max} at foveal vision, given in the previous chapter, and e is the eccentricity in degrees.

In section 2.4 of Chapter 2, it was already mentioned that the limit of the integration area formed by a maximum number of cycles is probably caused by the decrease of contrast sensitivity with increasing distance from the center of the retina. From measurements by Robson & Graham (1981, Fig. 3) with strip sized objects, it can be seen that for an eccentric object no limitation of the number of cycles occurs in tangential direction with respect to the center of the retina. Whereas for an object imaged in the center of the retina all dimensions are radial, for an object imaged outside the center, a distinction has to be made between radial and tangential dimensions. This means that for extra-foveal vision, the last term at the right-hand side of Eqs. (2.48) and (2.49) given in Chapter 2 has to be omitted for the tangential direction. For the radial direction, it should be taken into account that the edge of the integration area of an object in the center of the retina is located at a distance of $\frac{1}{2}N_{max}$ cycles from the center. This means that for an eccentric imaged object, N_{max} should not be used for the maximum number of cycles in radial direction, but $\frac{1}{2}N_{max}$. For an eccentric imaged object with radial size X and tangential size Y, therefore, the following modifications of Eqs. (2.48) and (2.49) have to be used:

$$X = \left(\frac{1}{X_o^2} + \frac{1}{X_{max}^2} + \frac{u^2}{(\frac{1}{2}N_{max})^2} \right)^{-0.5} \tag{4.19}$$

and

$$Y = \left(\frac{1}{Y_o^2} + \frac{1}{Y_{max}^2} \right)^{-0.5} \tag{4.20}$$

For eccentricities smaller than half the size of the object these formulae cannot be used. To arrive at expressions that can be used for all eccentricities, these equations can further be modified into

$$X = \left(\frac{1}{X_o^2} + \frac{1}{X_{max}^2} + \frac{(\frac{1}{2}X_o)^2 + 4e^2}{(\frac{1}{2}X_o)^2 + e^2} \frac{u^2}{N_{max}^2} \right)^{-0.5} \tag{4.21}$$

and

$$Y = \left(\frac{1}{Y_o^2} + \frac{1}{Y_{max}^2} + \frac{(\frac{1}{2}X_o)^2}{(\frac{1}{2}X_o)^2 + e^2} \frac{u^2}{N_{max}^2} \right)^{-0.5} \tag{4.22}$$

where by the introduction of an additional factor in the last term a smooth transition is obtained from centric to eccentric vision.

4.4 Comparison with measurements

With the formulae given in the preceding section, the constants of the foveal contrast sensitivity model given in section 3.8 of Chapter 3 can be adapted for extra-foveal vision. In this section published contrast sensitivity measurements made at extra-foveal vision will be compared with the so extended model. The measurements are given in chronological order of publication. The density of the on-center M-cells used in the calculations will be adapted for each eccentricity of the object to obtain a best fit with the high spatial frequency part of the measurements. The so obtained ganglion cell densities will be shown as a function of eccentricity in Fig. 4.17 at the end of this section in comparison with a curve for the average of the anatomical measurements by Curcio and Allen shown in Fig. 4.6. In the same way as in the preceding chapter, the foveal value of the constants σ_0, η, and k will be adapted to the measurements by a simultaneous fit for all eccentricities. This fit will be made by trial and error. If the contrast sensitivity is determined with the aid of a 2AFC method where the results do not correspond with 75% correct response, k will be corrected with Eq. (2.14) given in Chapter 2. A survey of the values used for σ_0, η, and k will be given in Table 4.1 at the end of this section.

4.4.1 Measurements by Virsu and Rovamo

In an investigation by Virsu & Rovamo (1979) that was already mentioned in section 3.9.10 of the previous chapter, they also made measurements of the contrast sensitivity function at different retinal eccentricities. The test object was a vertical or horizontal sinusoidal grating pattern generated on the screen of a high resolution monitor provided with a white phosphor (P4). The luminance of the test object was 10 cd/m^2 and the luminance of the surrounding area was the same. The shape of the

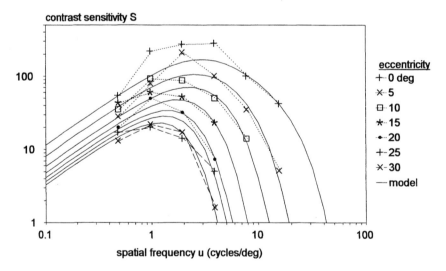

Figure 4.9: Contrast sensitivity function measured by Virsu & Rovamo (1979) at different eccentricities along the nasal half of the horizontal meridean of the retina. Luminance 10 cd/m^2. Circular field with a diameter of 5°. Monocular viewing with a natural pupil. The solid curves have been calculated with the extended model given in this chapter.

grating pattern used in one experiment was circular with a constant angular size of 5°. In this experiment, eccentricity was varied along the nasal half of the horizontal meridian of the retina. Fixation was made with both eyes, but the stimulus field was observed with only one eye (the right eye) with a natural pupil. The viewing distance was 2.28 m. The modulation threshold was determined with a 2AFC method where the threshold corresponded with 84% correct response. Seven subjects between 25 and 36 years of age participated in the experiments. The reported data are from one subject (VV). The measurements used here were measurements where the orientation of the grating had to be detected.

Measurements and calculations are shown in Fig. 4.9. The values of σ_0, η, and k used for the calculation were 0.48 arc min, 4.0%, and 4.3, respectively, after correcting k for the difference between 84% and 75% correct response with the aid of Eq. (2.14). Apart from some peaking at intermediate spatial frequencies, the general agreement between measurements and calculations is good.

In a second experiment, the eccentricity was varied along the superior half of the vertical meridian of the retina. In this experiment, the measurements were made with a semicircular grating of which the straight edge was opposing the fixation mark. The radius was 1° for zero eccentricity and increased with the eccentricity up to 15.8° in a part of the measurements. This variation was made by changing the viewing distance, starting with a viewing distance of 4.58 m for zero eccentricity. The

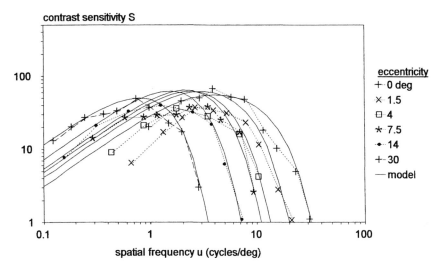

Figure 4.10: Contrast sensitivity function measured by Virsu & Rovamo (1979) at different eccentricities along the superior half of the vertical meridian of the retina. Luminance 10 cd/m^2. Semicircular field with a radius increasing with eccentricity. Monocular viewing with a natural pupil. The solid curves have been calculated with the extended model given in this chapter.

eccentricities mentioned by the authors refer to the position of the straight edge of the semicircular pattern that was nearest to the center of the retina. This has been taken into account in the calculations by adding half the effective radial field size to the mentioned eccentricity. Other conditions were the same as in the first experiment. The measurements used here are from observer PL for the situation where the field size varied with eccentricity.

Measurements and calculations are shown in Fig. 4.10. The values of σ_0, η, and k used for the calculation were 0.68 arc min, 3.0%, and 4.7, respectively, after correcting k for the difference between 84% and 75% correct response with the aid of Eq. (2.14). Compared with Fig. 4.8, the curves for the high eccentricities clearly show the effect of the simultaneously increased field size. This overcompensates the drop of contrast sensitivity at low spatial frequencies and means that the curves as a whole shift to lower spatial frequencies. Apart from some deviations at intermediate eccentricities, the calculated curves show a good agreement with the measurements.

4.4.2 Measurements by Robson and Graham

Robson & Graham (1981) measured the contrast sensitivity at different spatial frequencies as a function of eccentricity. The measurements were made along both halves of the vertical meridian of the retina. The test object was a horizontally oriented sinusoidal grating pattern generated on the screen of a CRT provided with P31 phosphor. The luminance of the test object was 500 cd/m^2 and the surrounding was kept on the same luminance. The test object contained four cycles and had a square shape without sharp edges. The spatial frequency on the display was fixed at 3 cycles/cm and the angular spatial frequency for the observer was varied by varying the viewing distance. For 3 cycles/deg, the viewing distance was 0.57 m. The eccentricity was expressed in the number of cycles of the concerning spatial frequency, as the authors wanted to investigate the effect of the number of cycles of the eccentricity on contrast sensitivity. The pattern on the display was turned on and off with a Gaussian temporal function lasting 167 msec above half its peak value. The test object was observed with both eyes and with a natural pupil. The observer had to fixate midway between two horizontally separated dark fixation marks. The modulation threshold was determined with a temporal 2AFC method where the threshold corresponded with 90% correct response. The authors served as subject. The reported data are from one subject: JR. Here, the average of the measurements for the two halves of the vertical meridian is used.

Measurements and calculations are shown in Fig. 4.11. The values of σ_0, η, and k used for the calculation were 0.67 arc min, 3.0%, and 4.9, respectively, after correcting k for the difference between 90% and 75% correct response with Eq. (2.14). The values of N_g have been calculated here with Eq. (4.9) using for e_g a

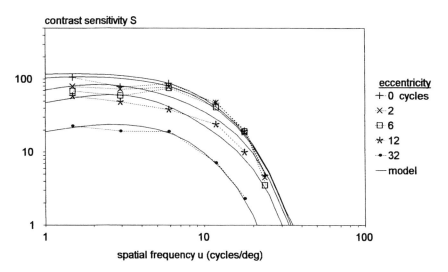

Figure 4.11: Contrast sensitivity function measured by Robson & Graham (1981) at different eccentricities along the vertical meridian of the retina. The eccentricity is varied with the number of cycles of the concerning spatial frequency and varies, therefore, along the curves (except for the curve at an eccentricity of zero cycles). Square test pattern consisting of 4 cycles. Luminance 500 cd/m². Binocular viewing with a natural pupil. The solid curves have been calculated with the extended model given in this chapter.

value of 3.9°. Note that the eccentricity is expressed in the number of cycles of the concerning spatial frequency and that the eccentricity, therefore, varies along the curves (except for the curve for an eccentricity of zero cycles). The somewhat flattened shape of the curves at small spatial frequencies is caused by the use of a fixed number of cycles in this experiment, similarly as the curve shown in Fig. 3.17 of Chapter 3. The fixed number of cycles reduces the angular field size of the object at increasing spatial frequency. Apart from a small dip of the measurement data at a spatial frequency of 3 cycles/deg, which is probably due to some measurement error, the measurements are in reasonably good agreement with the calculations.

Fig 4.12 shows the same measurements and calculations, but now plotted as a function of the eccentricity expressed in the number of cycles with the spatial frequency as parameter. From the figure, it can be seen that by plotting the measurements in this way, a nearly constant logarithmic decrease of contrast sensitivity with eccentricity is obtained, which is approximately equal for all spatial frequencies. This decrease is not influenced by limitations formed by the size of the object, as the test object contained only four cycles and its angular size was not larger than 2.7° for the lowest spatial frequency. In section 2.4 of Chapter 2, it has already been mentioned that the limitation of the integration area by a fixed maximum number of cycles is probably caused by the decrease of contrast sensitivity with increasing distance from the center of the retina. The approximately equal decrease of contrast sensitivity as

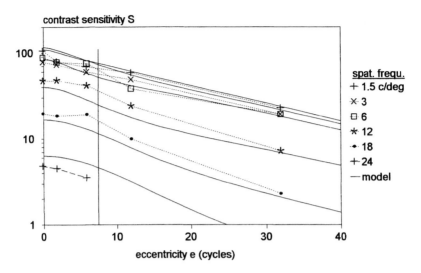

Figure 4.12: Same data as Fig. 4.11, but plotted as a function of the eccentricity expressed in the number of cycles of the concerning spatial frequency. The vertical line at an eccentricity of 7.5 cycles indicates the integration limit at foveal vision.

a function of the number of cycles shown in Fig. 4.12 supports the idea that the integration limit formed by a fixed maximum number of cycles is in fact caused by an equal decrease of contrast sensitivity for all spatial frequencies at an increase of the eccentricity with an equal number of cycles. At foveal vision, the limit of 15 cycles for the visual angle of an object corresponds with 7.5 cycles over its radial dimensions. This limit is shown in Fig. 4.12 by the vertical line at an eccentricity of 7.5 cycles. At this eccentricity, the contrast sensitivity has decreased with a factor of about 0.65 for all spatial frequencies. Areas with a lower contrast sensitivity probably do not contribute anymore to the integration process.

4.4.3 Measurements by Kelly

Kelly (1984) measured the contrast sensitivity function at different eccentricities using contiguous annular zones as stimuli. The stimuli consisted of radial grating patterns generated on the screen of a color CRT used in white mode. The annular zones had outside diameters of 4°, 8°, 16°, and 30°. The outside diameter of each zone coincided with the inner diameter of the neighbouring zone. The eccentricity was defined as the average of the inner and outer radius, except for the innermost area where the eccentricity was zero. The pattern was simultaneously varied in counter phase with a temporal sinusoidal variation of 0.5 Hz. This frequency may be assumed to be low enough to consider the measurements as static. See Chapter 5. The luminance was not mentioned, but was here assumed to be 10 cd/m^2 as most

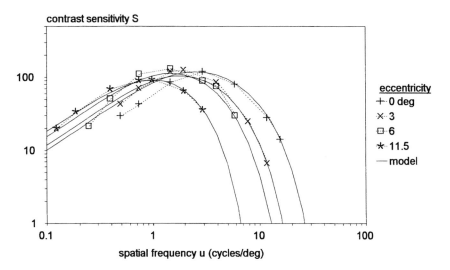

Figure 4.13: Contrast sensitivity function at different eccentricities measured by Kelly (1984) for neighbouring annular zones. Monocular viewing with an artificial pupil of 3 mm. The solid curves have been calculated with the extended model given in this chapter.

probable value. The object had a constant size and the eccentricity and the angular size of the pattern were, therefore, varied together by varying the viewing distance. The subject fixated a small dot in the center of the stimulus pattern and this pattern was stabilized on the retina by special equipment. Viewing was monocular with an artificial pupil of 3 mm. The modulation threshold was determined by the method of adjustment. The reported data are from one subject: the author.

Measurements and calculations are shown in Fig. 4.13. The values of σ_0, η, and k used for the calculation were 0.9 arc min, 2.5%, and 4.1, respectively. The general agreement between measurements and calculations is good.

4.4.4 Measurements by Mayer and Tyler

Mayer & Tyler (1986) measured the contrast sensitivity function at zero eccentricity and at an eccentricity of 3.5° along the inferior vertical meridian of the retina. The test object was a vertically oriented sinusoidal grating pattern generated on a monitor screen provided with P31 phosphor. The luminance was 40 cd/m². The angular width of the pattern was 4° and the angular height was 1°. The pattern was surrounded by a cardboard matched in color and luminance with the grating. Observers viewed the display with both eyes and with a natural pupil. The viewing distance was 1.52 m. The modulation threshold was determined by a Weibull function used as approximation of the psychometric function (See section 2.2 of Chapter 2). The quantity α of

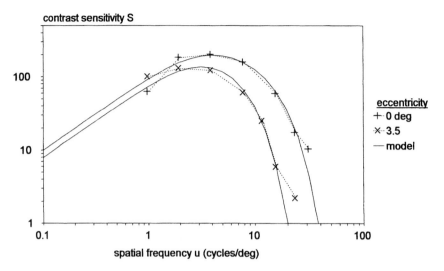

Figure 4.14: Contrast sensitivity function measured by Mayer and Tyler (1986) at zero eccentricity and at an eccentricity of 3.5° along the inferior half of the vertical meridian of the retina. Luminance 40 cd/m². Rectangular field with a width of 4° and a height of 1°. Binocular viewing with a natural pupil. The solid curves have been calculated with the extended model given in this chapter.

this function was used as modulation threshold. Three subjects participated in the experiments. Here, the data from one subject, the first author, are used, being the only subject for which foveal and extrafoveal data were reported.

Measurements and calculations are shown in Fig. 4.14. The values of σ_0, η, and k used for the calculation were 0.66 arc min, 2.2%, and 3.3, respectively, after correcting k for the quantity α of the Weibull function with the aid of Eqs. (2.7) and (2.10). The agreement between measurements and calculations is very good, apart from a strange crossing of the measured curves at a spatial frequency of about one cycle/deg and a strange deviation of these curves at the highest spatial frequency. These deviations are probably caused by some measurement error.

4.4.5 Measurements by Johnston

Johnston (1987) measured the contrast sensitivity function at different eccentricities in the nasal visual field. This field corresponds with the temporal half along the horizontal meridian of the retina. An emphasis was made on the high-frequency part of the contrast sensitivity curve. The test object was a vertical or horizontal sinusoidal grating pattern generated on the screen of an oscilloscope provided with blue-green P31 phosphor. The luminance was 10 cd/m². The surrounding of the pattern was masked with a white card with a luminance matched with the display. The pattern

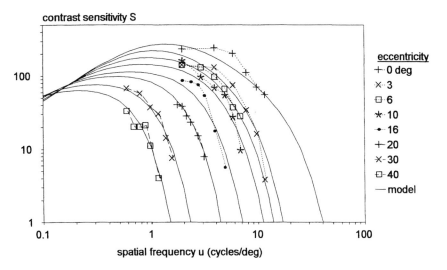

Figure 4.15: Contrast sensitivity function measured Johnston (1987) at different eccentricities along the temporal half of the horizontal meridian of the retina. Luminance 10 cd/m^2. Square field with a size inversely proportional to the spatial frequency. Monocular viewing with a natural pupil. The solid curves have been calculated with the extended model given in this chapter.

contained 12 cycles. It was square and had a constant size of 5.2 cm. The spatial frequency was varied by varying the viewing distance. In this way, the angular field size varied inversely proportionally with the spatial frequency. At 12 cycles/deg the viewing distance was 3 m and the angular size was 1°. Fixation was made with a fixation spot. Viewing was monocular with a natural pupil. The modulation threshold was determined with the aid of a temporal 2AFC method. The data used here are from one subject, the author, for vertically oriented gratings.

Measurements and calculations are shown in Fig. 4.15. The values of σ_0, η, and k used for the calculation were 0.5 arc min, 3.5%, and 3.4, respectively. The agreement between measurements and calculations is good.

4.4.6 Measurements by Pointer and Hess

Pointer and Hess (1989) measured the contrast sensitivity function at different eccentricities along the horizontal and vertical meridian of the retina with special emphasis on the low spatial frequency range. The test object was a horizontally oriented sinusoidal grating pattern generated on the screen of a CRT monitor provided with a white phosphor (P4). The luminance was 100 cd/m^2. The display screen was surrounded by a field with the same luminance. The pattern was circular with a circularly-symmetric two-dimensional Gaussian envelope and a Gaussian time

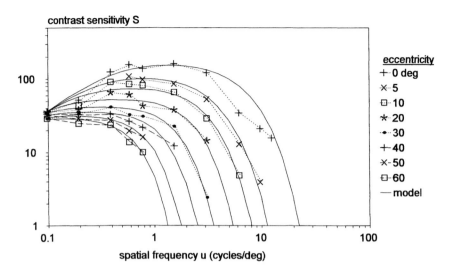

Figure 4.16: Contrast sensitivity function measured Pointer & Hess (1989) at different eccentricities along the horizontal meridian of the retina. Luminance 100 cd/m². Circular field with a size inversely proportional with spatial frequency. Monocular viewing with a natural pupil. The solid curves have been calculated with the extended model given in this chapter.

profile. The spatial and temporal Gaussian windows were truncated at their 1/e value. The spatial size was 6.4 cycles and the temporal duration was 500 msec. The spatial frequency was varied by varying the viewing distance between 0.18 and 3.7 m. In this way, the angular field size varied inversely proportionally with the spatial frequency. The spatial luminance patterns were presented with a temporal frequency of 1 Hz in counter phase. This frequency may be assumed to be low enough to consider the measurements as static. See Chapter 5. Fixation was made with a fixation light outside the center of the display. The test object was always in the center of the display. Viewing was monocular with the observer's dominant eye and with a natural pupil. The modulation threshold was determined with the aid of a temporal 2AFC method. Three subjects participated in the experiments. The data used here are the measurement results for the horizontal meridian from two subjects (PAB and JSP, the first author) averaged over subjects and over both halves of the meridian.

Measurements and calculations are shown in Fig. 4.16. The values of σ_0, η, and k used for the calculation were 1.1 arc min, 3.0%, and 4.1, respectively. Apart from the curve at zero eccentricity, the calculated curves are in very good agreement with the measurements. The shape of the curves in this figure and in Fig. 4.15 show a resemblance with a bird's head with a sharp beak. This is caused by the increase of the angular field size at low spatial frequencies. In Fig. 4.15 no measurements are available in this part of the curves. However, the measurements for this area shown in Fig. 4.16 confirm very well the predictions by the model.

4.4.7 Survey of the measurements

For the given measurements, the densities of the on-center M-cells used in the calculations were adapted for each eccentricity to obtain a best fit with the high spatial frequency part of the measurements. The so obtained densities are shown in Fig. 4.17 as a function of eccentricity. The continuous curve in this figure shows the average density of on-center M-cells derived from the anatomical measurements by Curcio and Allen. For this curve the approximation formula shown in Fig. 4.6 is used, after multiplication of the density with 0.05 based on the assumption that 5% of the total number of ganglion cells are on-center M-cells. Because of the spread of the measurement data, it appeared to make no sense to compare the data separately with the measurements for each of the concerning areas of the retina. The figure shows a general agreement between the ganglion cell density derived from the various contrast sensitivity measurements. There is also a reasonable agreement with the density of the on-center M-cells obtained from the anatomical data. However, at higher eccentricities, the density obtained from these data is generally lower than that derived from the contrast sensitivity measurements. This could mean that the assumption that 5% of the ganglion cells are on-center M-cells is not correct, and that this percentage is in fact closer to 7% or 10% at higher eccentricities. However, it could also mean that the ganglion cell density measured by Curcio and Allen is too low at high eccentricities. Recent measurements by Sjöstrand et al. along the vertical meridian show a much higher density at high eccentricities. See Sjöstrand et al. (1999, p. 2995).

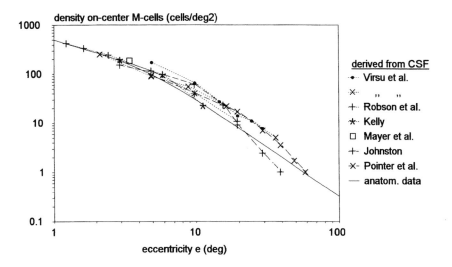

Figure 4.17: Density of on-center M-cells as a function of retinal eccentricity, derived from the contrast sensitivity measurements evaluated in this chapter. The solid curve represents the average of the anatomical data measured by Curcio & Allen (1990), assuming that 5% of the measured ganglion cells are on-center M-cells.

A survey of the values of σ_0, η, and k used for the evaluation of the measurements is given in Table 4.1. The values for η are close to 3%, but the values for σ_0 and k are generally somewhat higher than the data given in Table 3.1 of Chapter 3.

Table 4.1: σ_0, η, and k values used for the evaluation of the measurements

author	σ_0 (arc min)	η (%)	k
Virsu & Rovamo (1979)	0.48 0.68	4.0 3.0	4.3 4.7
Robson & Graham (1981)	0.67	3.0	4.9
Kelly (1984)	0.9	2.5	4.4
Mayer & Tyler (1986)	0.66	2.2	3.3
Johnston (1987)	0.5	3.5	3.4
Pointer & Hess (1989)	1.1	3.0	4.1

4.5 Summary and conclusions

In this chapter, the spatial contrast sensitivity model given in the previous chapter has been extended to extra-foveal vision. For this purpose, relations have been developed for the dependence of the constants used in the model on retinal eccentricity. For these relations, use has been made of biological and anatomical data of the density distribution of different retinal cell types. From this analysis, it appeared that especially the density variation of the on-center M ganglion cells has a large effect on extra-foveal contrast sensitivity. With the aid of the given relations, it is even possible to derive the density of these cells at different eccentricities from measurements of the contrast sensitivity function at these eccentricities.

The so extended model appeared to show a good agreement with various published measurements of extra-foveal contrast sensitivity. This was, of course, also partly due to many assumptions made in the model, which were based on results obtained with these data. However, the agreement for different types of measurements shows that these assumptions probably have a more general validity.

The density of the on-center M-cells that were derived from the contrast sensitivity measurements with the aid of the given equations appeared largely to agree with measurements of anatomical data by Curcio & Allen (1990). However, at larger eccentricities, the calculated densities were somewhat higher than the densities derived from the anatomical data. This could mean that the assumption that 5% of the ganglion cells are on-center M-cells has to be corrected to a higher value at high eccentricities. It could also mean that the measurements by Curcio and Allen give a too low density at these eccentricities. Sjöstrand et al. (1999) found recently with anatomical measurements higher densities at these eccentricities.

Measurements by Robson & Graham (1981) confirmed the presumption, made in section 2.4 of Chapter 2, that the limit of the integration area by a fixed maximum number of cycles is probably caused by a decrease of contrast sensitivity with increasing eccentricity. These measurements showed that this decrease is approximately equal for all spatial frequencies at an increase of the eccentricity with an equal number of cycles. From other measurements by Robson and Graham, it appeared that this limit occurs only in radial direction. At foveal vision all dimensions are radial so that this limit is always present, but at extra-foveal vision a distinction has to be made between radial and tangential dimensions of the retinal image with respect to the center of the retina.

References

Coletta, N.J. & Williams, D.R. (1987). Psychophysical estimate of extrafoveal cone spacing. *Journal of the Optical Society of America A*, **4**, 1503-1513.

Curcio, C.A. & Allen, K.A. (1990). Topography of ganglion cells in human retina. *Journal of Comparative Neurology*, **300**, 5-25.

Curcio, C.A., Sloan, K.R., Packer, O., Hendrickson, A.E., and Kalina, R.E. (1987). Distribution of cones in human and monkey retina: individual variability and radial asymmetry. *Science*, **236**, 579-582.

Henry, G.H. & Vidyasagar, T.R. (1991). Evaluation of mammalian visual pathways. In: J.R. Cronly-Dillon & R.L. Gregory (Eds.) *Vision and visual dysfunction, II. Evolution of the eye and visual system*, Chapter 20. MacMillan Press, London.

Johnston, A. (1987). Spatial scaling of central and peripheral contrast-sensitivity functions. *Journal of the Optical Society of America A*, **4**, 1583-1593.

Kelly, D.H. (1984). Retinal inhomogeneity. I. Spatiotemporal contrast sensitivity. *Journal of the Optical Society of America A*, **1**, 107-113.

Lee, B.B. (1996). Receptive field structure in the primate retina. *Vision Research*, **36**, 631-644.

Mayer, M.J. & Tyler, C.W. (1986). Invariance of the slope of the psychometric function with spatial summation. *Journal of the Optical Society of America A*, **3**,

1166-1172.

Østerberg, G. (1935). Topography of the layer of rods and cones in the human retina. *Acta Ophthalmologica. Suppl.*, **6**, 1-103.

Pointer, J.S. & Hess, R.F. (1989). The contrast sensitivity gradient across the human visual field: with emphasis on the low spatial frequency range. *Vision Research*, **29**, 1133-1151.

Polyak, S. (1957). The vertebrate visual system. University of Chicago Press, Chicago.

Robson, J.G. & Graham, N. (1981). Probability summation and regional variation in contrast sensitivity across the visual field. *Vision Research*, **21**, 409-418.

Sjöstrand, J., Olsson, V., Popovic Z., and Conradi, N. (1999). Quantitative estimations of foveal and extra-foveal retinal circuitry in humans. *Vision Research*, **39**, 2987-2998.

Virsu, V. & Rovamo, J. (1979). Visual resolution, contrast sensitivity, and the cortical magnification factor. *Experimental Brain Research*, **37**, 475-494.

Wässle, H., Grünert, U., Röhrenbeck, J., and Boycott, B.B. (1990). Retinal ganglion cell density and cortical magnification factor in the primate. *Vision Research*, **30**, 1897-1911.

Williams, D.R., (1988) Topography of the foveal cone mosaic in the living human eye. *Vision Research*, **28**, 433-454.

Chapter 5

Extension of the contrast sensitivity model to the temporal domain

5.1 Introduction

The contrast sensitivity model given in Chapter 3 was restricted to spatial contrast sensitivity. In this chapter the model will be extended to the temporal domain so that a spatiotemporal model is obtained that can also be used for purely temporal luminance variations. Temporal contrast sensitivity has already been intensively studied in the fifties by de Lange (1952, 1954, 1957, 1958a, 1958b, 1961) and in the sixties and seventies by Kelly (1960, 1961, 1971, 1972, 1979) and by Roufs (1972a, 1972b, 1973, 1974a, 1974b, 1974c). Kelly (1960) proposed to combine spatial and temporal contrast sensitivity measurements by using spatiotemporal stimuli to get more insight into spatiotemporal interactions. The two-dimensional contrast sensitivity function obtained with these types of stimuli is not simply the product of a spatial and a temporal response, but shows a much more complicated behavior (Robson, 1966; van Nes et al., 1967; Kelly, 1971, 1972, 1979; Koenderink & van Doorn, 1979). To explain this behavior, Kulikowski & Tolhurst (1973) supposed the existence of a sustained and a transient channel in the human visual system, analogous to supposed spatial frequency channels in the spatial domain.

For the temporal contrast sensitivity, it may be assumed that it is determined by internal noise in the same way as for spatial contrast sensitivity. Besides a spatial character, noise generally also has a temporal character. For the spatiotemporal model that will be given here, no different channels will be assumed for a sustained and a transient response, but the complicated spatiotemporal behavior of the visual system will be explained in a much simpler way. In this model, it is only assumed that the inhibition signal undergoes, in the inhibition process, besides spatial filtering, also temporal filtering before being subtracted from the photo-receptor signal.

This model has some similarity with a model proposed by Burbeck & Kelly (1980), where the spatiotemporal contrast sensitivity function is obtained as the difference between an excitatory mechanism and an inhibitory mechanism, and where each of these two responses is the product of an exclusively spatial and an exclusively temporal response curve. The shape of these response curves has to be determined by data fitting. However, our model is different from this model and has more resemblance with a model given by Fleet et al. (1985) where the inhibition signal is subtracted from the photo-receptor signal in a similar way as in the model of Burbeck and Kelly, but where different spatial and temporal functions are used. Also their treatment is different from the treatment that will be used here. Another approach was followed by Watson (1986) in a working model for temporal contrast sensitivity where two temporal response functions are subtracted from each other. However, this model, which is also different from our model, does not contain the spatial aspects and it can, therefore, be used only for temporal contrast sensitivity. Our model is based on a few simple assumptions about the temporal effects of the lateral inhibition.

5.2 Generalization of the contrast sensitivity model

The basic assumption of our model is, that the inhibition signal undergoes temporal filtering in addition to spatial filtering before being subtracted from the photo-receptor signal (Barten, 1993). For this purpose, Eq. (3.26) given in Chapter 3 for the spatial contrast sensitivity is modified by introducing a modification of the denominator of the term with Φ_0, which represents the squared MTF of the inhibition process. As mentioned in section 3.6 of Chapter 3, the MTF of the lateral inhibition process consists of a highpass filter given by $1 - F(u)$, where $F(u)$ is the MTF of the lowpass filter used for the inhibition. The shape of this function is described by Eq. (3.21). For the extension of the spatial contrast sensitivity model to the temporal domain is now assumed that the function $1 - F(u)$ has to be replaced by the function

$$G(u,w) = H_1(w)\{1 - H_2(w)F(u)\} \qquad (5.1)$$

where w is the temporal frequency, $H_1(w)$ is the MTF that represents the temporal filtering of the photo-receptor signal on its way from the photo-receptors to the brain and $H_2(w)$ is the MTF that represents the temporal filtering of the spatial inhibition signal before being subtracted from the photo-receptor signal. Fig 5.1 is a block diagram of the so obtained processing of information and noise. It is an extension of the block diagram shown in Fig. 3.1.

By replacing the denominator of the term with Φ_0 in Eq. (3.26) by $G^2(u,w)$ one obtains as equation for the here proposed spatiotemporal contrast sensitivity function

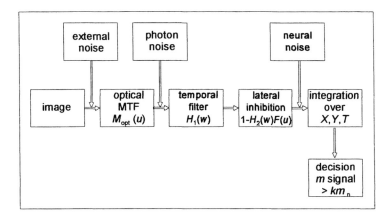

Figure 5.1: Block diagram of the processing of information and noise in the visual system according to the spatiotemporal contrast sensitivity model that is given here.

$$S(u,w) = \frac{M_{opt}(u)/k}{\sqrt{\frac{2}{T}\left(\frac{1}{X_o^2} + \frac{1}{X_{max}^2} + \frac{u^2}{N_{max}^2}\right)\left(\frac{1}{\eta p E} + \frac{\Phi_0}{[H_1(w)\{1 - H_2(w)F(u)\}]^2}\right)}} \tag{5.2}$$

where $F(u)$ is given by Eq. (3.21). This equation holds for equal image dimensions in x and y direction. For monocular instead of binocular vision, the factor 2 under the square-root sign has to be replaced by 4. The temporal functions $H_1(w)$ and $H_2(w)$ will be discussed in more detail in the following section.

5.3 Temporal filter functions

The functions $H_1(w)$ and $H_2(w)$ are temporal filter functions that describe the MTF of the temporal impulse response of the eye. According to de Lange (1952) the temporal filtering in the eye can be compared with the filtering by an electrical circuit of several cascaded filters that each consist of a combination of a resistance and a capacitance which is called an *RC filter*. The impulse response of one stage of such a filter is

$$h(t) = \frac{1}{\tau} e^{-t/\tau} \tag{5.3}$$

where $h(t) = 0$ for $t < 0$ and τ is the RC time. For a cascade of n such stages with the same RC time, the impulse response is

$$h(t) = \frac{n}{n!\,\tau} \left(\frac{t}{\tau} \right)^{n-1} e^{-t/\tau} \tag{5.4}$$

The MTF resulting from this impulse response can be obtained by taking the absolute value of its Fourier transform. This gives

$$H(w) = \frac{1}{\{1 + (2\pi w \tau)^2\}^{n/2}} \tag{5.5}$$

See, for instance, Papoulis (1968, p. 67). It appears that the functions $H_1(w)$ and $H_2(w)$ can indeed be described by this function with values τ_1, n_1 and τ_2, n_2, respectively, for the constants τ and n. This does not necessarily mean that the concerning impulse response is given by Eq. (5.4). As $H_1(w)$ and $H_2(w)$ contain only amplitude information and no phase information, the concerning impulse response functions cannot be found by an inverse Fourier transform of these functions. The temporal impulse response of the visual system will further be treated in section 5.9.

For $w\tau \gg 1$ the function given by Eq. (5.5) has an asymptotic slope $-n$ on double logarithmic scale. From Eq. (5.2), it follows then that contrast sensitivity as a function of temporal frequency will have a decay with a slope $-n_1$ on double logarithmic scale at high temporal frequencies. Daly and Normann (1985) measured the electrical response of cones in the eyes of turtles and found a decay corresponding with an asymptotic slope of about -6. They also mention that similar results have been found by other authors for other animals. Watson (1986) uses a slope -9 in his model. Roufs (1972b) already remarked that the value of n is not very critical for a fit with measured data, because of the spread of the data points, but that the value of τ depends on the chosen value of n. We found a best fit with measured data with a slope -7, so that a value 7 will be used for n_1 in our model. For n_2, a value 4 will be used, but a different value would also be possible, as most measurements are not very critical for this value.

The values of n_1 and n_2 are essentially fixed, because they are determined by the biological structure of the neural cells that transport the information, or by the number of synapses of these cells. Conversely, the time constants τ_1 and τ_2 depend on retinal illuminance and field size and can also be different for different subjects. In the following sections, a comparison of the model will be given with published contrast sensitivity measurements. For measurements with a series of data curves, a simultaneous fit will be made for all curves similarly as in the preceding chapters.

5.4 Spatiotemporal contrast sensitivity measurements

The spatiotemporal stimulus used in spatiotemporal contrast sensitivity measurements can generally be written in the form

$$L(x,t) = \overline{L}\{1 + m\cos(2\pi ux)\cos(2\pi wt)\} \qquad (5.6)$$

where $\overline{L}$ is the average luminance, and u and w are the spatial and temporal frequency, respectively. Besides this standing wave pattern, a traveling wave pattern is sometimes used. This has the form

$$L(x,t) = \overline{L}[1 + m\cos\{2\pi u(x - ct)\}] \qquad (5.7)$$

where c is the velocity of the pattern. With $w = u \times c$ and the use of a well-known trigonometric relation, Eq. (5.6) can be written as

$$L(x,t) = \overline{L}[1 + \tfrac{1}{2}m\cos\{2\pi u(x + ct)\} + \tfrac{1}{2}m\cos\{2\pi u(x - ct)\}] \qquad (5.8)$$

This means that a standing wave can be considered as the sum of two traveling waves moving in opposite directions. Therefore, eye movements can form a problem in measuring spatiotemporal contrast sensitivity. Kelly (1979) showed that with special measures for eye stabilization, no difference in contrast sensitivity is found between standing waves and traveling waves.

The first measurements of the combined spatiotemporal contrast sensitivity were made by Robson (1966). He measured the spatial contrast sensitivity function at four different temporal frequencies and the temporal contrast sensitivity function at four different spatial frequencies. The measurements were made at a luminance of 20 cd/m^2 using a vertically oriented sinusoidal grating pattern generated on the screen of an oscilloscope tube. The angular size of the test object was 2.5°×2.5° and the measurements were made at a viewing distance of 1 m. The observer looked at the test object with both eyes and with a natural pupil. The author (JGR), was the observer. The spatial contrast sensitivity measurements at the lowest temporal frequency (1 Hz) have already been mentioned in section 3.9.3 of Chapter 3, where it was assumed that this frequency is low enough to consider the measurements as static.

Fig. 5.2 shows the measurements of the spatial contrast sensitivity function at different temporal frequencies, while Fig. 5.3 shows the measurements of the temporal contrast sensitivity function at different spatial frequencies. The solid curves in both figures have been calculated with Eq. (5.2). The values used for τ_1 and τ_2 were 10.1 msec and 11.8 msec, respectively, and the values used for σ_0, η, and k were the same as used in section 3.9.3 of Chapter 3, being 0.53 arc min, 2.0% and 4.5, respectively. The calculated curves were simultaneously fitted with the combined series of data of both figures. This could explain the worse fit at higher spatial frequencies in Fig 5.3. Apart from this, the model gives a good description of the trends shown by the measurements.

From Fig. 5.2, it can be seen that the spatial contrast sensitivity function has its normal bandpass shape at low temporal frequencies, whereas it gets a lowpass

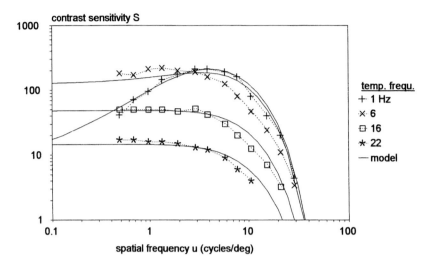

Figure 5.2: Spatial contrast sensitivity function measured by Robson (1966) at four different temporal frequencies. Luminance 20 cd/m². Field size 2.5°×2.5°. Binocular viewing a natural pupil. The solid curves have been calculated with Eq. (5.2).

shape at high temporal frequencies. According to our model, this can simply be explained by the difference in the temporal behavior of the photo-receptor signal and the subtracted lateral inhibition signal. At low temporal frequencies, $H_1 = H_2 \approx 1$, so $H_1\{1-H_2F(u)\} \approx 1-F(u)$, and the contrast sensitivity function shows its normal spatial

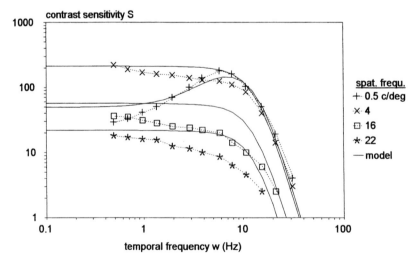

Figure 5.3: Temporal contrast sensitivity function measured by Robson (1966) at four different spatial frequencies. Luminance 20 cd/m². Field size 2.5°×2.5°. Binocular viewing with a natural pupil. The solid curves have been calculated with Eq. (5.2).

frequency behavior. At medium temporal frequencies H_2 is low, so $H_1\{1-H_2F(u)\} \approx H_1$. This means that the lateral inhibition has disappeared and that the contrast sensitivity curve is flat at low spatial frequencies. At higher temporal frequencies the shape remains the same but the level of the contrast sensitivity function further decreases proportionally to H_1.

Fig. 5.3 shows that the temporal contrast sensitivity function has a bandpass shape at low spatial frequencies and a lowpass shape at high spatial frequencies. According to the model given here, the bandpass character at low spatial frequencies is caused by a gradual decrease of the lateral inhibition when the temporal frequency increases. At low spatial frequencies, $F(u) \approx 1$, so $H_1\{1-H_2F(u)\} \approx H_1(1-H_2)$. The factor $1-H_2$ causes a reduction of the contrast sensitivity at low temporal frequencies in the otherwise flat part of $H_1(w)$. For high spatial frequencies, $F(u) \approx 0$, so $H_1\{1-H_2F(u)\} \approx H_1$, which means that the contrast sensitivity curve is flat at low temporal frequencies. All curves start at low temporal frequencies from a value determined by the spatial contrast sensitivity function. The resemblance in shape between the curves of this figure and that of Fig. 5.2 is remarkable.

Kelly (1979) also made measurements of the spatial contrast sensitivity function at different temporal frequencies. They are shown in Fig. 5.4. He used a circular test pattern with a diameter of 7.5° with a vertically oriented sinusoidal grating pattern generated on the screen of a CRT monitor. Viewing was monocular with an artificial pupil of 2.3 mm and with the use of special equipment to stabilize

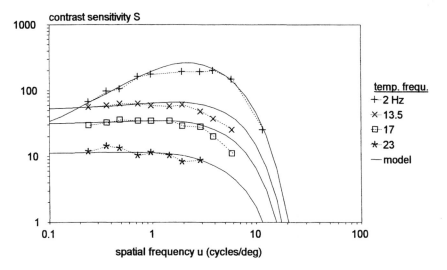

Figure 5.4: Spatial contrast sensitivity function measured by Kelly (1979) at four different temporal frequencies. Retinal illuminance 300 Td. Circular field with a diameter of 7.5°. Monocular viewing with an artificial pupil of 2.3 mm. The solid curves have been calculated with Eq. (5.2).

the position of the object on the retina. The retinal illuminance was 300 Td. The modulation threshold was determined by the method of adjustment. The measurements were made by one subject. The solid curves in the figure have been calculated with Eq. (5.2). The values used for τ_1 and τ_2 were 11.1 msec and 5.0 msec, respectively, and the values used for σ_0, η, and k were 1.35 arc min, 1% and 3.2, respectively. The curves have the same shape as the curves in Fig. 5.2. They show a good agreement with the measurements.

5.5 Temporal contrast sensitivity measurements

Temporal contrast sensitivity measurements are usually not made with a spatial grating pattern, but with an evenly illuminated uniform field. From the general contrast sensitivity function given by Eq. (5.2), the spatial contrast sensitivity can be obtained by inserting $w = 0$. However, the temporal contrast sensitivity function cannot be obtained by simply inserting $u = 0$ in this equation. This insertion is correct for the factor $M_{lat}(u)$, which becomes 1, but is not correct regarding the function $F(u)$. The reason is that spatial field dimensions still play a role in temporal contrast sensitivity, because they determine the amount of lateral inhibition. For a uniform field the fundamental wave given by the size of the field is the strongest spatial frequency component of the object. The spatial frequency of this wave may therefore be considered to represent the spatial frequency content of the object in the function $F(u)$. If the field is square and has an angular width X_o, the spatial frequency of the fundamental wave is

$$u = \frac{1}{2X_o} \tag{5.9}$$

If X_o is expressed in degrees, the spatial frequency u is given in cycles/deg. For a circular field, it may be assumed that the effective spatial frequency is equal to that of a square field with the same surface area. If the angular diameter of the field is D, the spatial frequency is then given by

$$u = \frac{1}{\sqrt{\pi} D} \tag{5.10}$$

The spatial frequency given by one of these equations has to be inserted in $F(u)$, instead of $u=0$, to obtain a correct expression for the temporal contrast sensitivity from Eq. (5.2).

That the spatial content of a uniform field can well be described by this spatial frequency is shown in Fig. 5.5, where two types of temporal contrast sensitivity measurements made by Kelly (1971) are given. One type was made with a circular uniform field with a diameter of 7° and the other type was made with a grating

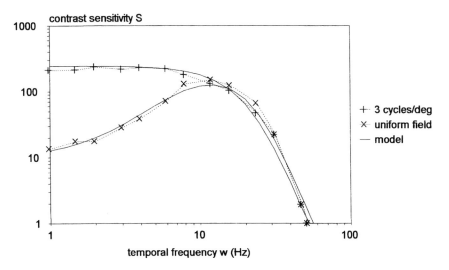

Figure 5.5: Temporal contrast sensitivity measured by Kelly (1971) for a grating with a spatial frequency of 3 cycles/deg and for a uniform field. Both fields are circular with a diameter of 7°. Retinal illuminance 1670 Td. Monocular viewing with an artificial pupil of 2.3 mm. The solid curves have been calculated with Eq. (5.2) using for the uniform field the value of u given by Eq. (5.10).

pattern of the same size with a spatial frequency of 3 cycles/deg. The test objects were generated on the screen of a CRT monitor. They were observed with the right eye at a distance of 0.5 m through an artificial pupil of 2.3 mm. The retinal illuminance was 1670 Td. The modulation threshold was determined by the method of adjustment. The subject was a female (LH), 20 years of age. The solid curves in the figure have been calculated with Eq. (5.2) where for the uniform field, the fundamental spatial frequency given by Eq. (5.10) was used for $F(u)$ and M_{spat} was set to 1. The values used for τ_1 and τ_2 were 6.9 msec and 6.1 msec, respectively, and the values used for σ_0, η, and k were 0.5 arc min, 3% and 5.6, respectively. The simultaneous fit of both curves with the measurement data shows, that the temporal contrast sensitivity of a uniform field can well be described by Eq. (5.2) by using the spatial frequency of the fundamental wave in $F(u)$.

Another example of temporal contrast sensitivity measurements for a uniform field is given in Fig. 5.6. This figure shows measurements by Roufs & Blommaert (1981) for a circular field with a diameter of 1° and a retinal illumination of 1200 Td. Viewing was monocular with an artificial pupil of 2 mm. The modulation threshold was determined by measuring the psychometric function and using the modulation at 50% detection probability. Subject was the first author (JAJR), 46 years of age. The solid curve was calculated in the same way as for the uniform field in Fig. 5.5. The values used for τ_1 and τ_2 were 7.2 msec and 13.6 msec, respectively, and the values used for η and k were 3% and 3.9, respectively. The curve is very similar to the

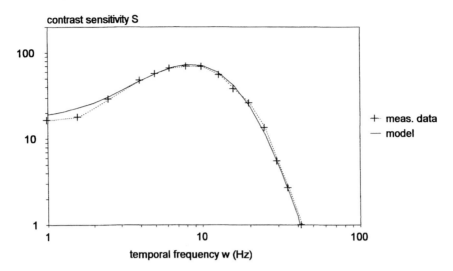

Figure 5.6: Temporal contrast sensitivity measured by Roufs and Blommaert (1981) for a circular uniform field with a diameter of 1° and a retinal illuminance of 1200 Td. Monocular viewing with an artificial pupil of 2 mm. The solid curve has been calculated with Eq. (5.2) using the value of u given by Eq. (5.10).

curve for the uniform field measured by Kelly, which was shown in Fig. 5.5. The fit between calculations and measurements is very good.

In the uniform field, higher harmonics of the fundamental spatial frequency can also play a role in the detection process. In the frequency spectrum, only odd harmonics are present. For small fields, the spatial frequencies of the harmonics fall in the declining part of the spatial contrast sensitivity function and can therefore be neglected. For large fields, the fundamental spatial frequency is low and the spatial frequency of the higher harmonics falls partly in the rising part of the spatial contrast sensitivity function. This is illustrated by contrast sensitivity measurements made by Campbell & Robson (1968) for sine-wave and square wave gratings (See, for instance, their Fig. 3). Therefore, with large fields generally a better fit with the measurements is obtained by using for the spatial frequency in $F(u)$ the third harmonic, instead of the fundamental spatial frequency.

The spatial frequency used in $F(u)$ has only an influence on the contrast sensitivity at low temporal frequencies. At temporal frequencies above 10 Hz, $F(u)$ has practically no influence because $H_2(w)$ becomes too small. This can be seen from Fig. 5.5 where the curves above 10 Hz nearly coincide.

5.6 Effect of a surrounding field

Most temporal contrast sensitivity measurements are made with a dark area surrounding the test field. However, sometimes a stationary surrounding field is used with a constant luminance equal to the average luminance of the test field. This method was, for instance, used by de Lange in his measurements. Roufs (1972a) pointed out that this detail may not be neglected. With a surrounding field the temporal contrast sensitivity is increased. The effect of the surround can be taken into account in the calculations by assuming that the effective size of the object is increased, whereas the fundamental spatial frequency to be used in $F(u)$ remains the same.

To investigate the effect of a surrounding field, Roufs (1972a) measured the temporal contrast sensitivity of a uniform field with and without an equiluminous surrounding field. The test field was a circular field with a diameter of 1° and a retinal illuminance of 1150 Td. Viewing was monocular with an artificial pupil of 2 mm. The threshold was determined by measuring the psychometric function and using the modulation at 50% detection probability. The author, JAJR, was the observer, 39 years of age at the time of this investigation.

Measurements and calculations are shown in Fig. 5.7. The values used for τ_1

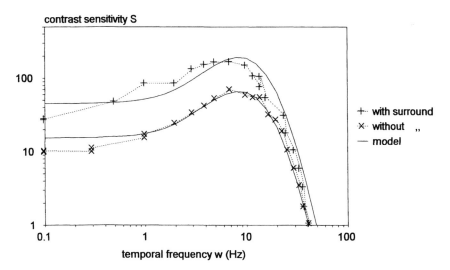

Figure 5.7: Temporal contrast sensitivity measured by Roufs (1972a) for a circular uniform field with a diameter of 1° with and without a stationary equiluminous surround. Retinal illuminance 1150 Td. Monocular viewing with an artificial pupil of 2 mm. The solid curves have been calculated with Eq. (5.2), using for u the value given by Eq. (5.10) and using for the measurements with surrounding field an effective field size with a threefold diameter.

and τ_2 were 7.3 msec and 13.2 msec, respectively, and the values used for η and k were 3% and 4.2, respectively. Both solid curves have been calculated with Eq. (5.2) using for u the value given by Eq. (5.10). However, for the curve with a surrounding field, an integration area with a threefold diameter was used to obtain an agreement with the measurements. This size increase corresponds with the addition of an annular ring with a width equal to the diameter of the stimulus. From the figure, it can be seen that in this way a good description of the measurements is obtained. This rule will therefore also be applied here in other situations where the uniform field is surrounded by an equiluminous stationary field. At the very low frequencies used in this experiment, one can observe some deviation between measurements and calculations. However, one should realize that the measurement of the contrast sensitivity at low temporal frequencies is very difficult. At a temporal frequency of 0.1 Hz, one cycle lasts 10 sec. It is well known that it is very difficult for a subject to observe luminance chances that happen so slowly.

5.7 Effect of retinal illuminance and field size on the time constants

Temporal contrast sensitivity measurements made at different retinal illuminances cannot be described with a single value for the time constants τ_1 and τ_2. The same holds for measurements made at different field sizes. The time constants appear to depend on both quantities. They decrease with increasing retinal illuminance and with increasing field size. The dependence on field size is probably related to the variation of cone density and ganglion cell density over the retina. By analyzing the published temporal contrast sensitivity measurements given in this section, it was found that the dependence on retinal illuminance and field size can approximately be described by the following equations:

$$\tau_1 = \frac{\tau_{10}}{1 + 0.55 \ln \left\{ 1 + \left(1 + \dfrac{D}{1} \right)^{0.6} \dfrac{E}{3.5} \right\}} \tag{5.11}$$

and

$$\tau_2 = \frac{\tau_{20}}{1 + 0.37 \ln \left\{ 1 + \left(1 + \dfrac{D}{3.2} \right)^{5} \dfrac{E}{120} \right\}} \tag{5.12}$$

where τ_{10} and τ_{20} are fixed time constants that do not depend on retinal illuminance and field size, D is the field diameter in degrees, and E is the retinal illuminance in Troland.

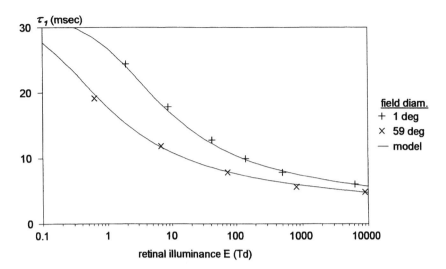

Figure 5.8: Variation of the time constant τ_1 with retinal illuminance, calculated with Eq. (5.11) for a circular field with a diameter of 1° and 59°. The data points for a diameter of 1°are derived from measurements by Roufs (1972a) and the data points for a diameter of 59° are derived from measurements by Kelly (1961).

Figs. 5.8 and 5.9 show how τ_1 and τ_2, respectively, vary with retinal illuminance according to these formulas. Curves are given for two field diameters: 1° and 59° with measured data points derived from temporal contrast sensitivity measure-

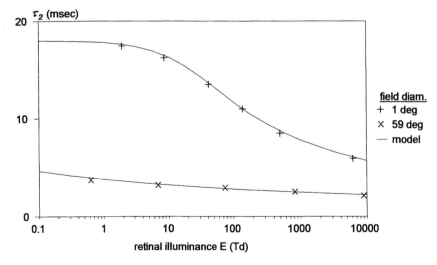

Figure 5.9: Variation of the time constant τ_2 with retinal illuminance, calculated with Eq. (5.12) for a circular field with a diameter of 1° and 59°. The data points for a diameter of 1°are derived from measurements by Roufs (1972a) and the data points for a diameter of 59° are derived from measurements by Kelly (1961).

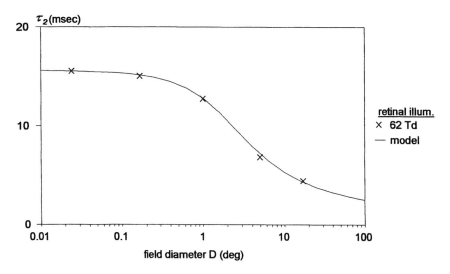

Figure 5.10: Variation of the time constant τ_2 with the diameter of a circular field, calculated with Eq. (5.12) for a retinal illuminance of 62 Td. The data points are derived from measurements by Roufs & Bouma (1980).

ments. The data for 1° were derived from measurements by Roufs (1972a) and the data for 59° were derived from measurements by Kelly (1961). These measurements will be treated in more detail in the following part of this section. From the figures, it can be seen that the field size mainly has an influence on τ_2. The dependence of τ_2 on field size is further shown in Fig. 5.10 for a fixed retinal illuminance of 62 Td. The data points in this figure were derived from measurements by Roufs & Bouma (1980). These measurements will also be treated in more detail in the following part of this section.

At very low retinal illuminance and very small field size, the time constants τ_1 and τ_2 become equal to τ_{10} and τ_{20}, respectively. The actual values of τ_1 and τ_2 are characterized by these constants. Ideally τ_{10} and τ_{20} have the same value for all measurements, but in practice, they differ for different subjects and for different experiments. In Figs. 5.8 to 5.10, $\tau_{10} = 32$ msec and $\tau_{20} = 18$ msec. These values can be considered as typical values for these constants. Eqs. (5.11) and (5.12) will be used in the following temporal contrast sensitivity measurements for a simultaneous description of the curves for different field sizes or illuminances. For these measurements, values of τ_{10} and τ_{20} will be used that give a best fit with the data. The results will be given in Table 1 at the end of this section.

Fig. 5.11 shows so obtained calculated curves with the measurement data for temporal contrast sensitivity measurements made by Roufs & Bouma (1980) for a large range of field sizes. The data given in Fig. 5.10 were obtained from this

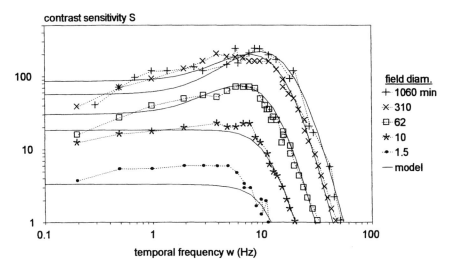

Figure 5.11: Temporal contrast sensitivity measured by Roufs & Bouma (1980) for circular fields with different field diameters. Retinal illuminance 62 Td. Monocular viewing with an artificial pupil of 2 mm. The solid curves have been calculated with Eq. (5.2) and Eqs. (5.10) through (5.12).

investigation. The measurements were made with a retinal illuminance of 62 Td and with circular fields of which the diameter was varied from 1.5 arc min (0.025°) to 1060 arc min (17.7°). Viewing was monocular with the left eye through an artificial pupil of 2 mm. Subject was HJM, 25 years of age. The solid curves have been calculated with Eq. (5.2) and Eqs. (5.10) through (5.12) with a simultaneous fit for all curves. The values of τ_2 shown in Fig. 5.10 were obtained from the same data by a non-simultaneous fit. For the two largest field sizes in the figure the third harmonic of the spatial frequency was used in $F(u)$, instead of the fundamental frequency, for the reasons mentioned in section 5.5. The values of τ_{10} and τ_{20} were 32 msec and 18 msec, respectively, and the values used for η and k were 6% and 2.1, respectively. Apart from a deviation for the very small field diameter of 1.5 arc min (0.025°), measurements and calculations show a very good agreement over a large range of field sizes. This means that the dependence of the time constants on field size is well described by Eqs. (5.11) and (5.12).

Fig. 5.12 shows measurements by de Lange (1958a) for a large range of retinal illuminance levels extending from 0.375 Td to 1000 Td. He used a circular test field with a diameter of 2° surrounded by a uniform field with a diameter of 60° at the same luminance. Viewing was monocular with an artificial pupil of 2.8 mm. The modulation threshold was determined by the method of adjustment. Two subjects took part in the experiments. In Fig. 5.12, the data are given for one subject, subject L, the author, 52 years of age. The solid curves in the figure have been calculated with Eq. (5.2) and Eqs. (5.10) through (5.12). The presence of the surrounding field

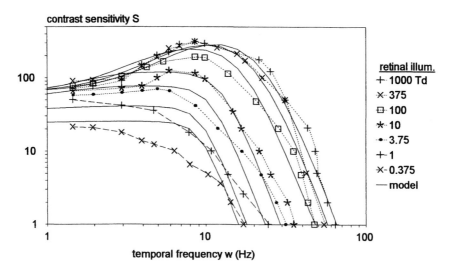

Figure 5.12: Temporal contrast sensitivity curves measured by de Lange (1958a) for a circular field with a diameter of 2° and a large range of retinal illuminance levels. Monocular viewing with an artificial pupil of 2.8 mm. The solid curves have been calculated with Eq. (5.2) and Eqs. (5.10) through (5.12).

was taken into account in the calculations by increasing the diameter of the integration area with a factor 3, as mentioned in section 5.6. The values of τ_{10} and τ_{20} were 29 msec and 18 msec, respectively, and the values used for η and k were 2% and 2.8, respectively. Apart from some considerable deviations, the general trend of the measurements is well described by the calculated curves. The deviations can partly be explained by the primitive conditions under which these first temporal contrast sensitivity measurements had to be made. The deviations of the two lowest curves can be explained by the fact that the retinal illuminance for these curves is scotopic, instead of photopic.

After the investigation by de Lange, other investigators (e.g., Kelly and Roufs) studied intensively the dependence of temporal contrast sensitivity on luminance. Because of the pioneering work by de Lange, temporal contrast sensitivity curves are often called *de Lange curves*.

Fig. 5.13 shows measurements by Kelly (1961) for a large circular test field with a smooth edge and a range of retinal illuminance levels extending from 0.65 Td to 9300 Td. The 50% diameter of the test field was 59°. Viewing was monocular with an artificial pupil of 1.55 mm. The modulation threshold was determined by the method of adjustment. The author, DHK, was the observer. The solid curves in the figure have been calculated with Eq. (5.2) and Eqs. (5.10) through (5.12) with a simultaneous fit for all curves. The data for 59° given in Figs. 5.8 and 5.9 were obtained from this investigation, but the values of τ_1 and τ_2 shown in these figures

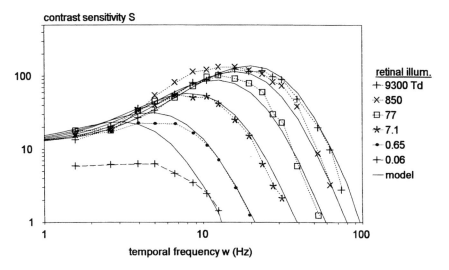

Figure 5.13: Temporal contrast sensitivity measured by Kelly (1961) for a circular field with a diameter of 59° and a large range of retinal illuminance levels. Monocular viewing with an artificial pupil of 1.55 mm. The solid curves have been calculated with Eq. (5.2) and Eqs. (5.10) through (5.12) using the third harmonic of the fundamental spatial frequency.

were obtained from a non-simultaneous fit. Because of the large field, the third harmonic of the fundamental spatial frequency was used for the spatial frequency in $F(u)$. See section 5.5. The values of τ_{10} and τ_{20} were 31 msec and 17 msec, respectively, and the values used for η and k were 3% and 3.6. Apart from the curve for 0.06 Td, where vision is scotopic, the agreement between measurements and calculations is very good.

Fig. 5.14 shows similar measurements by Roufs (1972a) for a circular test field with a diameter of 1° and a range of retinal illuminance levels extending from 2 Td to 6500 Td. Viewing was monocular with an artificial pupil of 2 mm. The modulation threshold was determined by measuring the psychometric function and using the modulation at 50% detection probability. Subject was RK, 18 years of age. The solid curves in the figure have been calculated with Eq. (5.2) and Eqs. (5.10) through (5.12) with a simultaneous fit for all curves. The data points for 1° given in Figs. 5.8 and 5.9 were obtained from these measurements, but the values of τ_1 and τ_2 shown in these figures were obtained from a non-simultaneous fit. The values of τ_{10} and τ_{20} were 33 msec and 18 msec, respectively, and the values used for η and k were 2% and 2.6, respectively. From the figure, it can be seen that the general agreement between measurements and calculations is very good.

Fig. 5.15 shows other measurements by Roufs (1973) for three retinal illuminance levels: 1 Td, 42 Td, and 1200 Td. The measurements were made under

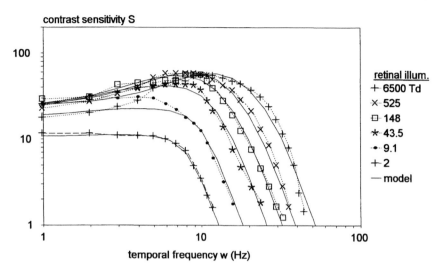

Figure 5.14: Temporal contrast sensitivity curves measured by Roufs (1972a) for a circular field with a diameter of 1° and a large range of retinal illuminance levels. Monocular viewing with an artificial pupil of 2 mm. The solid curves have been calculated with Eq. (5.2) and Eqs. (5.10) through (5.12).

the same conditions as for the measurements described above, but with a different subject, JAJR, the author, 39 years of age at the time of this investigation. The solid curves in the figure have been calculated with Eq. (5.2) and Eqs. (5.10) through (5.12). The values of τ_{10} and τ_{20} were 32 msec and 17 msec, respectively, and the

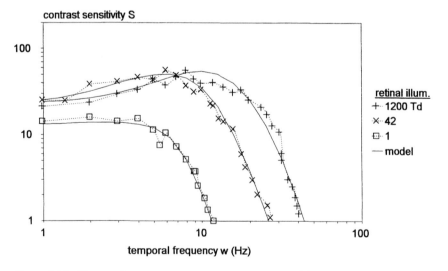

Figure 5.15: Temporal contrast sensitivity curves measured by Roufs (1973) for three different retinal illuminance levels. Apart from a different observer, further conditions are the same as in Fig. 5.14.

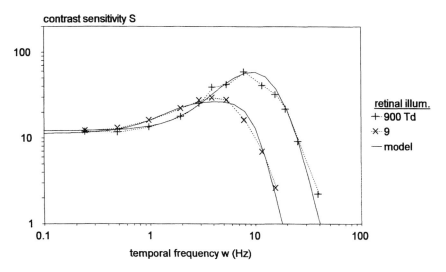

Figure 5.16: Temporal contrast sensitivity curves measured by Swanson et al. (1987) for a circular field with a diameter of 2° and two different retinal illuminance levels. Monocular viewing with an artificial pupil of 2 mm. The solid curves have been calculated with Eq. (5.2) and Eqs. (5.10) through (5.12).

values used for η and k were 8% and 2.8, respectively. The agreement between measurements and calculations is very good.

Fig. 5.16 shows measurements by Swanson et al. (1987) for a circular test field with a diameter of 2° and for two of the four measured retinal illuminance levels: 9 Td and 900 Td. Viewing was monocular with an artificial pupil of 2 mm. The modulation threshold was determined by the method of adjustment. Two subjects took part in the experiments: WS and TU, the first two authors. The given data are the average for these subjects. The solid curves have been calculated with Eq. (5.2) and Eqs. (5.10) through (5.12). The values of τ_{10} and τ_{20} were 35 msec and 23 msec, respectively, and the values used for η and k were 3% and 5.3, respectively. The calculated curves show a good agreement with the measurements.

From the evaluation of the measurement data given in this section, it appears that Eqs. (5.11) and (5.12) give a good description of the dependence of the time constants on retinal illuminance and field size. In Table 5.1 a survey is given of the values of τ_{10} and τ_{20} used for the evaluation of these measurements. Besides these values, values are also added that were derived from the measurements given in the preceding sections. From the table, it can be seen that τ_{10} and τ_{20} show a concentration around 32 msec, and 18 msec, respectively. These values can, therefore, be considered as typical values for these constants. Contrary to what one should expect, the data in the table do not show a clear effect of age. However, this corresponds with the results of an investigation by Tyler (1989) who measured temporal contrast sensitivity as a function of age for a large number of subjects. For subjects between 16 and 70 years of age, he found only a very small dependence on age.

Table 5.1: Time constants τ_{10} and τ_{20} used for the measurements

author	field size (deg)	retinal illum. (Td)	τ_{10} (msec)	τ_{20} (msec)	subject	age (years)
Robson (1966)	2.5	320	41	31	JGR	±26
Kelly (1979)	7.5	300	46	18	---	---
Kelly (1971)	7	1670	35	25	LH	20
Roufs et al. (1981)	1	1200	32	32	JAJR	46
Roufs (1972a)	1	1150	32	31	JAJR	39
Roufs et al. (1980)	varied	62	32	18	HJM	25
de Lange (1958)	2	varied	29	18	L	52
Kelly (1961)	59	varied	31	17	DHK	±25
Roufs (1972a)	1	varied	33	18	RK	18
Roufs (1973)	1	varied	32	17	JAJR	39
Swanson et al. (1987)	1	varied	35	23	av. 2	---

5.8 Flicker sensitivity: Ferry-Porter law

The study of temporal contrast sensitivity was stimulated by the introduction of motion picture films at the end of the nineteenth century and was stimulated again by the introduction of television halfway through the twentieth century. The reason for this interest was the annoying effect of flicker that can occur at viewing the reproduced pictures. This effect is caused by the repetition of images, which is needed for the simulation of movement, and is a function of the frequency of the repetition. Around the end of the nineteenth century, Ferry (1892) and Porter (1902) found that the frequency up to which flicker can be observed, increases linearly with the logarithm of the luminance. This law is known as Ferry-Porter law. This frequency is called the *critical flicker frequency* or CFF. At the time of these first investigations, measurements had to be made with very primitive means. For a part of his experiment, Porter used candles as light source. Later, when television started, the higher luminance of television images caused a new interest for the flicker problem. At this time, de Lange did his well-known investigation on the fundamental aspects of periodic temporal luminance variations.

The critical flicker frequency can be derived from the temporal contrast sensitivity function. From this function, the frequency can be calculated where the contrast sensitivity reaches a value 1. At this frequency, a modulation of 100% is needed to observe a luminance variation with a probability of 50% (or with a probability of 75% in a 2AFC experiment). Extensive measurements of the critical flicker frequency have been made by Tyler & Hamer (1990) who found an accurate match with the Ferry-Porter law over a large range of retinal illuminance levels. Fig. 5.17 shows their measurement results for a circular field with a diameter of 0.5° and 0.05° with a 100% modulated sinusoidal luminance variation. Viewing was monocular. The observer was RDH, the second author. The curves through the data points have been calculated with Eq. (5.2) and Eqs. (5.10) through (5.12) for the situation that $S = 1$. The values of τ_{10} and τ_{20} were 31 msec and 18 msec, respectively, and the values used for η and k were 3% and 3.0, respectively. These values are about equal to the typical values of these constants. The agreement between measurements and calculations is very good.

The given expressions can also be applied to calculate the critical flicker

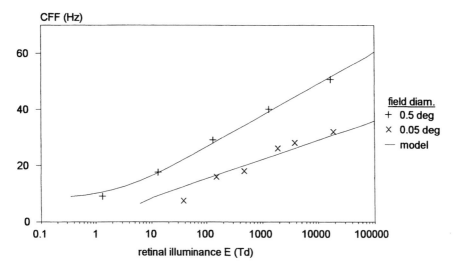

Figure 5.17: Critical flicker frequency as a function of retinal illuminance measured by Tyler & Hamer (1990) for a circular field with a diameter of 0.5° and 0.05° with a 100% modulated sinusoidal temporal luminance variation. Viewing was monocular. The solid curves have been calculated with Eq. (5.2) and Eqs. (5.10) through (5.12) for $S = 1$ using a simultaneous fit for both curves.

frequency for the practical situation of images displayed on a *cathode ray tube* or CRT used for televison or computer display. In this situation, viewing is binocular and the luminance variation is not sinusoidal, but consists of an exponential decay with a repetition frequency equal to the frame rate. For a non-sinusoidal luminance variation, the first harmonic of the variation has to be used to determine the visibility of flicker. The decay time of the exponential decay is usually short with respect to the repetition time of the signal. For this situation, the amplitude of the first harmonic is nearly twice the average luminance. However, such a situation would only occur if the light emission would start simultaneously in all the points of the image. In practice, this is not so, because the light emission starts successively in the different points of the image. This successive emission causes an effectively flat dependence of the total image luminance on time, except for the complete darkness of the image during the vertical retracing. By the interruption during the vertical retracing, the actual time dependence of the luminance obtains a rectangular shape. Fig. 5.18 shows a sketch of this idealized rectangular luminance variation. The first harmonic or fundamental wave of this pattern is the cause for the visibility of flicker. The modulation of this sinusoidal wave can be calculated with the aid of a Fourier analysis. For a rectangular luminance variation holds

$$m = \frac{2\sin(\alpha\pi)}{\alpha\pi} \tag{5.13}$$

where m is the modulation of the first harmonic, and α is the relative time part of active luminance. For television and computer display, the vertical ray tracing takes generally about 8% of the frame time, so that α is 0.92. From Eq. (5.13) follows that

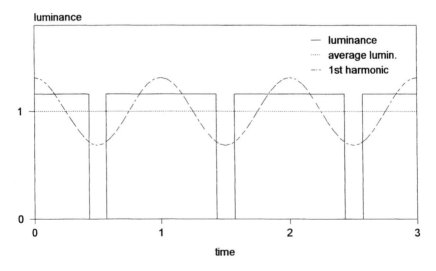

Figure 5.18: Solid curve: idealized temporal luminance variation of CRT images. Dotted curve: average luminance. Dashed curve: first harmonic of the temporal luminance variation.

the modulation of the fundamental wave is then 17.2%, which means that for S a value of $1/0.172 = 5.81$ has to be used to calculate the critical flicker frequency. With this value of S and with the aid of Eq. (5.2) and Eqs. (5.10) through (5.12), the critical flicker frequency has been calculated as a function of the luminance for a circular field with a diameter of $30°$. This field size represents the average viewing condition for computer displays. For τ_{10} and τ_{20} the values 32 msec and 18 msec, respectively, were used and for η and k the values 3% and 3.0, respectively, being the typical values of these constants. Besides the normal situation of 50% correct response, also a calculation was made for 10% correct response, which corresponds with a probability of 90% for not seeing flicker. This calculation was made by using a k value 1.72, instead of 3.0, as can be calculated for this situation with the aid of Eqs. (2.2) through (2.4) given in Chapter 2. The results are shown in Fig. 5.19. This figure also shows measurements by Farrell et al. (1987) of the 90% flicker limit for a CRT display seen with a subtended angle of $30°$. This limit corresponds with a probability of 90% for not seeing flicker. The data were derived by Farrell from the mean and the standard deviation of the flicker thresholds observed by 20 observers. These data appear to correspond very well with the calculations. The slope of the curve for this situation is about 14 Hz per decade. From the figure, it can further be seen that for a luminance level of 100 cd/m², flicker is still visible with a probability of 50% at a frame rate of 62 Hz. At a frame rate of 68 Hz, the probability of seeing flicker is reduced to 10%. From practical experience, it is known that the choice of a frame rate of 72 Hz for computer displays is sufficient to avoid flicker completely.

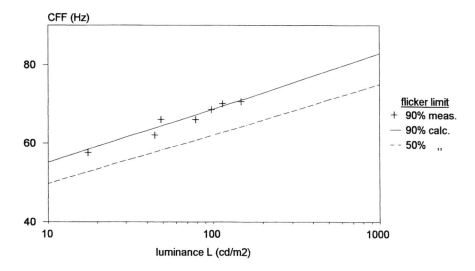

Figure 5.19: Critical flicker frequency as a function of the luminance for a CRT image seen with a subtended angle of $30°$. Data points: CFF measurements by Farell et al. (1987) of the 90% flicker limit that corresponds with a chance of 10% for seeing flicker. Solid curve: 90% limit calculated with our model. Dashed curve: same calculation for 50% probability of seeing flicker.

5.9 Temporal impulse response

In the previous sections, only the temporal frequency effects of the extended contrast sensitivity model were treated. The temporal impulse response function given by the model can be derived from the following equation that corresponds with Eq. (5.1):

$$h(t) = h_1(t) \star \{1 - F(u) h_2(t)\} \qquad (5.14)$$

where $h_1(t)$ is the impulse response of the temporal processing that the signal undergoes on its way from the photo-receptors to the brain, $h_2(t)$ is the impulse response of the temporal processing of the inhibition signal, $F(u)$ is the MTF of the lowpass filter of the lateral inhibition process given by Eq. (3.21), and the symbol $\star$ denotes convolution. This equation can also be written in the form

$$h(t) = h_1(t) - F(u)\{h_1(t) \star h_2(t)\} \qquad (5.15)$$

The functions $h_1(t)$ and $h_2(t)$ are the inverse Fourier transforms of the complex functions of which the functions $H_1(w)$ and $H_2(w)$, respectively, represent the absolute value. As was already mentioned in section 5.3, the functions $H_1(w)$ and $H_2(w)$ contain only amplitude information and not the required phase information, so that the impulse response function cannot be found by an inverse Fourier transform of these functions. However, we will here assume that the impulse response of these functions is simply given by Eq. (5.4), so that the total temporal impulse response can be calculated by using this equation for $h_1(t)$ and $h_2(t)$ in Eq. (5.15). For a short pulse, the so obtained impulse response function has a *triphasic shape*. This means that it starts with a negative part, followed by a positive part and ending again with a negative part, as is shown in Fig. 5.20. The negative part is caused by the second term in Eq. (5.15) which represents the lateral inhibition. The shape of this function is different from the *biphasic shape* consisting of a positive part followed by a negative part, which is generally assumed in other models. See Watson (1982).

At present, no biological measurements of the impulse response function are available. However, Roufs & Blommaert (1981) developed a sophisticated psycho-physical method to measure this function indirectly. They used a probe flash and a test flash with a very short duration compared with the duration of the impulse response. The duration of both flashes was 2 msec. Both flashes were superimposed on a constant luminance level. The test flash has a much smaller intensity than the probe flash and proceeds the probe flash, or is delayed with respect to it, by a variable time difference. At each time difference, the intensity of both flashes is varied with the same factor until the combined intensity of probe flash and test flash is just observed. The impulse response function is derived from the intensities of the flashes, the time difference between the flashes, and the threshold of the probe flash without test flash. This technique is similar to the technique developed by the same authors for a measurement of the spatial point-spread function, which was mentioned in section 3.6 of Chapter 3.

Fig. 5.20 shows the measurement results given by Blommaert & Roufs (1987) for subject JAJR, the second author. The measurements were made with a circular field with a diameter of 1° and a retinal illuminance of 1200 Td. The data points are the average of two measurement series. As the method gives no information about the zero point of the time scale, the time scale was arbitrarily set to zero at the maximum response. Furthermore, the response was arbitrarily set to 1 at this maximum. The solid curve in the figure has been calculated with Eqs. (5.15) and (5.4) with $\tau_1 =$ 7.2ms and $\tau_2 = 13.5$ msec, being the values that were obtained with the measurements of Fig. 5.6, which were made with the same subject and under the same conditions regarding field size and illuminance. $F(u)$ has been calculated with Eqs. (3.21) and (5.10). This gives for these measurements a value of 0.93. The agreement of the calculated curve with the measurements shows that the use of Eq. (5.4) for the calculation of the impulse response function is indeed correct and that the use of this function together with Eq. (5.15) gives a good description of the temporal response function of the eye. Note that in Fig. 5.20 the same values for the time constants are used as in Fig. 5.6. It must, however, be remarked that the calculations show a disagreement with the measurements in the first negative part of the curve. The total area under the negative part of the temporal response function must theoretically be nearly equal to the positive part. For the measurements, the area under the negative part is much larger then follows from the theoretical ratio $F(u) : 1 = 0.93 : 1$ used for the calculations. This means that the negative inhibition would be stronger than the positive excitation. As this is unlikely, it must be assumed that this deviation has probably some other cause.

Watson (1982) mentioned that the results of the measurements by Roufs and

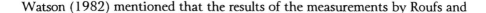

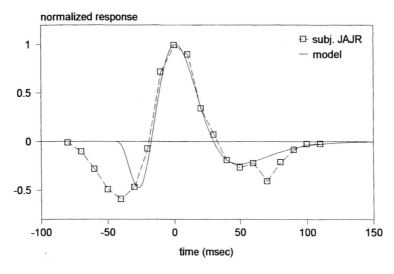

Figure 5.20: Temporal impulse response function measured by Blommaert and Roufs (1987). The solid curve has been calculated with Eqs. (5.18) and (5.7).

Blommaert need not be equal to the temporal response function and that a temporal response function with a biphasic shape could also explain their results. However, a triphasic shape was also found later by Tyler (1992) with a different measurement technique. It should further be remarked that the triphasic shape has a positive effect on the observation of temporal signals by the eye. With a triphasic shape, a negative part of the temporal response function precedes the main positive part. This gives a sharpening effect of temporal luminance changes, similar to the sharpening effect in the space domain known as Mach-band effect. Both effects are caused by lateral inhibition and both effects improve the possibility of the eye to observe luminance changes.

5.10 Summary and conclusions

In this chapter, the spatial contrast sensitivity model described in Chapter 3 has been extended to the temporal domain so that it can also be used for the temporal contrast sensitivity. The extension is based on the assumption that the lateral inhibition signal undergoes temporal filtering in addition to spatial filtering. With the so obtained spatiotemporal model, the remarkable spatiotemporal behavior of the visual system reported in several publications can simply be explained. Lateral inhibition appears to play an important role in these phenomena.

In the model two different time constants are used that both depend on retinal illuminance and field size. From an analysis of published temporal contrast sensitivity data, an approximation formula has been derived for the dependence of the time constants on these parameters. In this way the temporal contrast sensitivity of an individual observer can be characterized by two time constants that are independent of these parameters. The so obtained model appeared to be in very good agreement with published measurements.

The temporal contrast sensitivity model can also be used to calculate the critical flicker frequency for television and data display. The calculated results appeared to be in very good agreement with published measurements. The given method for the calculation of the flicker sensitivity can be used for a technical design of these systems.

The temporal contrast sensitivity model can also be used to calculate the temporal impulse response of the eye. With some additional assumption, the so obtained impulse response function appears to have a triphasic shape. This shape is different from the biphasic shape obtained with other models. Although no biological measurements of the temporal impulse response function are available, the prediction obtained with the model appeared to be in good agreement with psychophysical measurements by Roufs and Blommaert.

References

Barten, P.G.J. (1993). Spatio-temporal model for the contrast sensitivity of the human eye and its temporal aspects. *Human Vision, Visual Processing, and Digital Display IV, Proc. SPIE*, **1913**, 2-14.

Blommaert, F.J.J. & Roufs, J.A.J. (1987). Predictions of thresholds and latency on the basis of experimentally determined impulse responses. *Biological Cybernetics*, **56**, 329-344.

Burbeck, C.A. & Kelly, D.H. (1980). Spatiotemporal characteristics of visual mechanisms: excitatory-inhibitory model. *Journal of the Optical Society of America A*, **70**, 1121-1126.

Campbell, F.W. & Robson, J.G. (1968). Application of Fourier analysis to the visibility of gratings. *Journal of Physiology*, **197**, 551-566.

Daly, S.J. & Normann, R.A. (1985). Temporal information processing in cones: effects of light adaptation on temporal summation and modulation. *Vision Research*, **25**, 1197-1206.

de Lange, H. (1952). Experiments on flicker and some calculations on an electrical analogue of the foveal systems. *Physica*, **18**, 935-950.

de Lange, H. (1954). Relationship between critical flicker-frequency and a set of low-frequency characteristics of the eye. *Journal of the Optical Society of America*, **44**, 380-389.

de Lange, H. (1957). Attenuation characteristics and phase-shift characteristics of the human fovea-cortex systems in relation to flicker fusion phenomena. Ph. D. Thesis, Delft Technical University, Delft, The Netherlands.

de Lange, H. (1958a). Research into the dynamic nature of the human fovea-cortex systems with intermittent and modulated light. I. Attenuation characteristics with white and colored light. *Journal of the Optical Society of America*, **48**, 777-784.

de Lange, H. (1958b). Research into the dynamic nature of the human fovea-cortex systems with intermittent and modulated light. II. Phase shift in brightness and delay in color perception. *Journal of the Optical Society of America*, **48**, 784-789.

de Lange, H. (1961). Eye's response at flicker fusion to square-wave modulation of a test field surrounded by a large field of equal mean luminance. *Journal of the Optical Society of America*, **51**, 415-421.

Farrell, J.E., Benson, B.E., and Haynie, C.R. (1987). Predicting flicker thresholds for visual displays. *Proceedings of the SID*, **28**, 449-453.

Ferry, E.S. (1892). Persistence in vision. *American Journal of Science*, **44**, 192-207.

Fleet, D.J., Hallett, P.E., and Jepson, A.D. (1985). Spatiotemporal inseparability in early visual processing. *Biological Cybernetics*, **52**, 153-164.

Kelly, D.H. (1960). J_0 stimulus patterns for visual research. *Journal of the Optical Society of America*, **50**, 1115-1116.

Kelly, D.H. (1961). Visual responses to time-dependent stimuli. I. Amplitude sensitivity measurements. *Journal of the Optical Society of America*, **51**, 422-429.

Kelly, D.H. (1971). Theory of flicker and transient responses, II: counterphase gratings. *Journal of the Optical Society of America*, **61**, 632-640.

Kelly, D.H. (1972). Adaptation effects on spatio-temporal sine-wave thresholds. *Vision Research*, **12**, 89-101.

Kelly, D.H. (1979). Motion and vision, II: stabilized spatio-temporal threshold surface. *Journal of the Optical Society of America*, **69**, 1340-1349.

Koenderink, J.J. & van Doorn A.J. (1979). Spatiotemporal contrast detection threshold surface is bimodal. *Optics Letters*, **4**, 32-34.

Kulikowski, J.J. & Tolhurst, D.J. (1973). Psychophysical evidence for sustained and transient detectors in human vision. *Journal of Physiology*, **232**, 149-162.

Papoulis, A. (1968). Systems and transforms with applications in optics. McGraw-Hill, New York-St. Louis-San Francisco-Toronto-London-Sydney.

Porter, T.C. (1902). Contributions to the study of flicker. *Proceedings of the Royal Society, London A*, **70**, 313-329.

Robson, J.G. (1966). Spatial and temporal contrast-sensitivity functions of the visual system. *Journal of the Optical Society of America*, **56**, 11417-1142.

Roufs, J.A.J. (1972a). Dynamic properties of vision-I. Experimental relationships between flicker and flash thresholds. *Vision Research*, **12**, 261-278.

Roufs, J.A.J. (1972b). Dynamic properties of vision-II. Theoretical relationships between flicker and flash thresholds. *Vision Research*, **12**, 279-292.

Roufs, J.A.J. (1973). Dynamic properties of vision-III. Twin flashes, single flashes and flicker fusion. *Vision Research*, **13**, 309-323.

Roufs, J.A.J. (1974a). Dynamic properties of vision-IV. Thresholds of decremental flashes, incremental flashes and doublets in relation to flicker fusion. *Vision Research*, **14**, 831-852.

Roufs, J.A.J. (1974b). Dynamic properties of vision-V. Perception lag and reaction time in relation to flicker and flash thresholds. *Vision Research*, **14**, 853-869.

Roufs, J.A.J. (1974c). Dynamic properties of vision-IV. Thresholds of decremental flashes, incremental flashes and doublets in relation to flicker fusion. *Vision Research*, **14**, 831-852.

Roufs, J.A.J. & Blommaert, F.J.J. (1981). Temporal impulse and step responses of the human eye obtained psychophysically by means of a drift-correcting perturbation technique. *Vision Research*, **21**, 1203-1221.

Roufs, J.A.J. & Bouma, H. (1980). Towards linking perception research and image quality. *Proceedings SID*, **21**, 247-270.

Swanson, W.H., Ueno, T., Smith, V.C., and Pokorny, J. (1987). Temporal modulation sensitivity and pulse-detection thresholds for chromatic and luminance perturbations. *Journal of the Optical Society of America A*, **4**, 1992-2005.

Tyler, C.W. (1989). Two processes control variations in flicker sensitivity over the life span. *Journal of the Optical Society of America A*, **6**, 481-490.

Tyler, C.W. (1992). Psychophysical derivation of the impulse response through generation of ultrabrief responses: complex inverse estimation without minimum-phase assumptions. *Journal of the Optical Society of America A*, **9**, 1025-1040.

Tyler, C.W. & Hamer, R.D. (1990). Analysis of visual modulation sensitivity. IV. Validity of the Ferry-Porter law. *Journal of the Optical Society of America A*, **7**, 743-758.

van Nes, F.L., Koenderink, J.J., Nas, H., and Bouman, M.A. (1967). Spatiotemporal modulation transfer in the human eye. *Journal of the Optical Society of America*, **57**, 1082-1088.

Watson, A.B. (1982). Derivation of the impulse response: comments on the method of Roufs and Blommaert. *Vision Research*, **22**, 1335-1337.

Watson, A.B. (1986). Temporal sensitivity. In: K.R. Boff, L. Kaufman, and J.P. Thomas (Eds.) *Handbook of Perception and Human Performance, I. Sensory Processes and Perception*, chapter 6. Wiley, New York.

Chapter 6

Effect of nonwhite spatial noise
on contrast sensitivity

6.1 Introduction

In Chapter 2 where the effect of external noise on contrast sensitivity was treated, it was assumed that the noise was white. This means that the spectral density of the noise is constant within the frequency limits of the considered spectrum. Although these conditions are usually met, this is not always the case. Sometimes the spectral density is not constant within the frequency limits of the noise spectrum or is constant only within a limited part of this spectrum. In this chapter, the effect of nonwhite noise on contrast sensitivity will be treated. The disturbance of the observation of a signal by noise with a frequency that is different from that of the signal is called *masking*. It will be investigated how the formulae for the effect of white noise given in Chapter 2 can be generalized to also become valid for the situation of nonwhite noise. This treatment will, however, be restricted to spatial noise. If temporal noise is also present, it will be assumed that this noise is white.

6.2 Model for the masking effect of nonwhite spatial noise

For white noise, the threshold elevation of a signal by the presence of external noise is given by Eq. (2.50) in Chapter 2:

$$m_t' = \sqrt{m_t^2 + k^2 m_n^2}$$

where m_t is the modulation threshold without noise, m_t' is the increased modulation threshold with noise, and m_n is the average modulation of the noise wave components given by Eq. (2.43):

$$m_n = 2 \sqrt{\frac{\Phi_n}{XYT}}$$

For white noise, Φ_n and m_n are constant over the whole frequency spectrum of the noise band. For nonwhite noise, the spectral density of the noise will vary with the spatial frequency u_n of the noise, and m_n will therefore be different for different spatial frequencies u of the signal. To generalize the equations for white noise, so that they can also be used for nonwhite noise, we assume that m_n can be obtained by using the following equation (Barten, 1995):

$$m_n(u) = 2 \sqrt{\frac{\Phi_d(u)}{XYT}} \qquad (6.1)$$

This expression is similar as Eq. (2.43), but Φ_n has been replaced by the function $\Phi_d(u)$. This function describes the equivalent effect of the different spatial frequency components of the noise on the signal at spatial frequency u. For white noise, the function $\Phi_d(u)$ is simply equal to Φ_n. For nonwhite noise, we assume that $\Phi_d(u)$ can be derived from the frequency spectrum of the noise with the aid of the following relation:

$$\Phi_d(u) = \int_0^\infty \Psi(u_n, u) \, \Phi_n(u_n) \, \frac{du_n}{u} \qquad (6.2)$$

where u_n is the spatial frequency of the noise, and $\Psi(u_n, u)$ is a dimensionless weighting function that describes the masking effect of noise wave components with spatial frequency u_n on the observation of a signal with spatial frequency u.

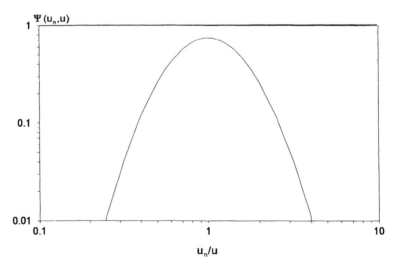

Figure 6.1: Plot on double logarithmic scale of the function $\Psi(u_n, u)$ given by Eq. (6.4) which describes the masking effect of noise components with spatial frequency u_n on a signal with spatial frequency u.

To obtain that $\Phi_d(u)$ is equal to Φ_n for white noise, the function $\Psi(u_n,u)$ has to meet the following condition:

$$\int_0^\infty \Psi(u_n,u) \frac{du_n}{u} = 1 \tag{6.3}$$

The function $\Psi(u_n,u)$ will generally be zero for most values of u_n and will differ from zero only for a range of values around u. Furthermore, $\Psi(u_n,u)$ appears to be a function of u_n/u. From measurements by Stromeyer & Julesz (1972) about masking effects by nonwhite noise, which will be treated in the next section, we derived the following empirical relation:

$$\Psi(u_n,u) = 0.747 \; e^{-2.2 \ln^2(u_n/u)} \tag{6.4}$$

This expression was obtained from the data by trial and error. The factor 0.747 has been chosen such that the condition of Eq. (6.3) is met. The function is shown in Fig. 6.1. It is a log normal distribution function, described by a Gaussian distribution of $\ln(u_n/u)$, with a 50% width of nearly two octaves.

A special situation of nonwhite noise occurs when the spectral density of the noise is constant, but the frequency band of the noise is outside the spatial frequency of the signal. This situation is shown in Fig. 6.2. Then Eq. (6.2) can be simplified to

$$\Phi_d(u) = \Phi_n \int_{u_{nmin}}^{u_{nmax}} \Psi(u_n,u) \frac{du_n}{u} \tag{6.5}$$

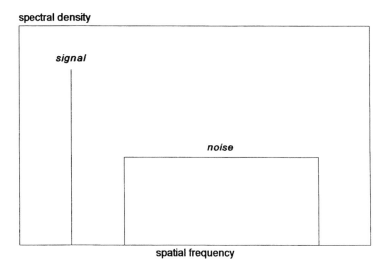

Figure 6.2: Noise spectrum for the situation of one-dimensional white noise in a spatial frequency band outside the spatial frequency of the signal.

where u_{nmin} and u_{nmax} are the minimum and maximum spatial frequency of the noise, respectively. Often the relative standard deviation σ_n of the noise is given, instead of the spectral density. Then Eq. (2.41) of Chapter 2 can be used to calculate Φ_n.

6.3 Measurements with narrow noise bands by Stromeyer and Julesz

Stromeyer & Julesz (1972) measured the threshold elevation of a large series of spatial frequencies using a one-dimensional vertically oriented dynamic noise pattern of which the spatial frequency range of the noise band was varied. The spectral density was constant within the noise band. The edges of the bands had a steepness of 42 dB/octave and their position was determined by measuring the -3 dB points. Noise intensity was determined by measuring σ_n. The stimuli were vertically oriented sinusoidal gratings displayed on a high-resolution monitor provided with a white phosphor (P4). The frame rate was 60 Hz and the average luminance was 15.9 cd/m². In most of the experiments, the viewing distance was 4 m and the stimulus field was 2.5°×1°. The test field was surrounded by a dark area. Viewing was binocular with a natural pupil. The modulation threshold was determined by the method of adjustment. Two subjects out of three (MHW, RAP, and CFS, the first author) served as observer in different parts of the experiments. For the evaluation of the data, the average of the results of these two observers was used to reduce the experimental spread.

For the one-dimensional dynamic noise used in these experiments, Eq. (2.41) of Chapter 2 gives

$$\Phi_n = \frac{\sigma_n^2}{2(u_{nmax} - u_{nmin})\, 2w_{nmax}} \tag{6.6}$$

so that Eq. (6.5) becomes

$$\Phi_d(u) = \frac{\sigma_n^2}{2(u_{nmax} - u_{nmin})\, 2w_{nmax}} \int_{u_{nmin}}^{u_{nmax}} \Psi(u_n, u)\, \frac{du_n}{u} \tag{6.7}$$

For one-dimensional dynamic noise Eq. (6.1) becomes

$$m_n(u) = 2 \sqrt{\frac{\Phi_d(u)}{XT}} \tag{6.8}$$

The authors expressed the measurement results as a relative threshold elevation defined by

$$l = \frac{m_t'}{m_t} - 1 \tag{6.9}$$

From this relation follows with the aid of Eq. (2.50) of Chapter 2

$$l = \sqrt{\frac{m_n^2(u)}{(m_t/k)^2} + 1} - 1 \tag{6.10}$$

In this expression, m_t/k can be calculated with Eq. (3.26). The actual value of k is irrelevant, because m_t is proportional with k.

With these equations, the relative threshold elevation l was calculated for the measurement conditions of the experiment, using Eq. (6.4) for the function $\Psi(u_n, u)$. Figs. 6.3 and 6.4 show the measurement data and the calculations for five different one-octave wide noise bands. The measurements for the lowest noise band in Fig. 6.3 were made at a viewing distance of 0.5 m with a field size of $20° \times 8°$, whereas the measurements for the lowest noise band in Fig. 6.4 were made at a viewing distance of 1 m with a field size of $10° \times 4°$. The other measurements in both figures were made at a viewing distance of 4 m with a field size of $2.5° \times 1°$. For most of the measurements the value of σ_n was 0.059 except the lowest and the highest noise bands in Fig. 6.3 for which σ_n was 0.042 and 0.047, respectively. Subjects MHW and CFS were the observers. The average of their measurement results was used in the figures. The values of σ_0 and η used for the calculation were 0.5 arc min and 4%, respectively.

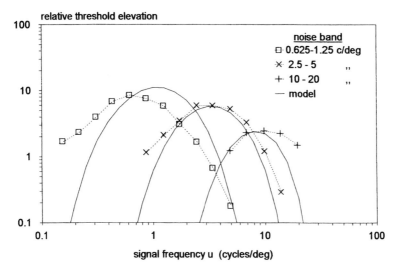

Figure 6.3: Relative threshold elevation by one-octave wide noise bands measured by Stromeyer & Julesz (1972). Luminance 15.9 cd/m². Binocular viewing with a natural pupil. The solid curves have been calculated with Eqs. (6.7) through (6.10) and Eqs. (3.26) and (6.4).

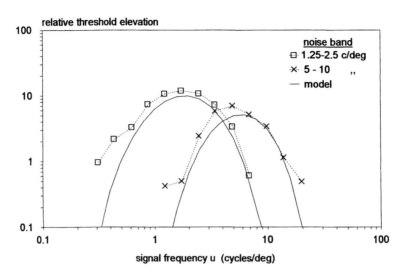

Figure 6.4: Same as Fig. 6.3 for two other one-octave wide noise bands.

Apart from measurements with narrow noise bands, Stromeyer and Julesz also made measurements with wide noise bands of variable size. Fig. 6.5 shows the threshold elevation by lowpass noise as a function of the maximum spatial frequency of the noise for three different spatial frequencies of the test signal. In this experiment

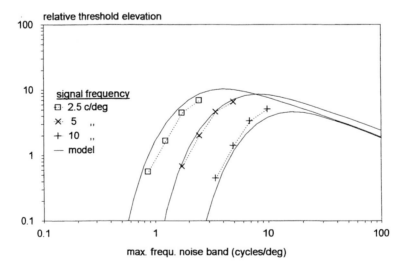

Figure 6.5: Relative threshold elevation by lowpass noise measured by Stromeyer & Julesz (1972) as a function of the maximum frequency of the noise band for three different frequencies of the test signal. Luminance 5.19 cd/m². Binocular viewing with a natural pupil. The solid curves have been calculated with Eqs. (6.7) through (6.10) and Eqs. (3.26) and (6.4).

σ_n was maintained at a fixed value of 0.15 during the variation of the maximum frequency of the noise. The constant value of σ_n causes a decrease of the spectral noise density with increasing maximum frequency of the noise. This explains the decrease of the threshold elevation at the right-hand side of the curves. Subjects RAP and CFS were the observers. The average of their measurement results has been used in the figure. The solid curves have been calculated in the same way as for the measurements shown in Figs. 6.3 and 6.4. The values of σ_0 and η used for the calculation were 0.5 arc min and 1.2%, respectively. The calculated curves in Figs. 6.3 through 6.5 show that Eq. (6.4) gives on the average a good description of the function $\Psi(u_n,u)$.

6.4 Measurements with nonwhite noise by van Meeteren and Valeton

To test if Eq. (6.4) is also valid for other measurements, an analysis has been made of measurements by van Meeteren & Valeton (1988). They measured the contrast sensitivity function without noise and with three types of nonwhite two-dimensional static noise that differed in bandwidth. Vertically oriented sinusoidal grating patterns with and without noise were generated on the screen of a video monitor. The patterns had a luminance of 100 cd/m^2 and were surrounded by a large field with the same luminance. They were observed from a distance of 3.5 m. The field size was $1° \times 1°$. Viewing was binocular with a natural pupil. Two subjects with normal visual acuity (AVM and MV, the authors) took part in the experiments. The modulation threshold was determined by measuring the psychometric function. The displayed picture consisted of 180×180 pixels. Noise was generated with a computer by assigning a random value from a uniform luminance distribution to each pixel. In this way fine-grained noise was made. Apart from this fine-grained noise, medium-grained noise and coarse-grained noise was generated by assigning random luminance values to a rectangular grid of pixels. The spacing of this grid was 5 pixels and 20 pixels, respectively, whereas the luminance value of the remaining pixels was interpolated by using a first-order Bessel function. In this way, the maximum spatial frequencies of the noise bands were 90 cycles/deg, 18 cycles/deg, and 4.5 cycles/deg in both directions for the three types of noise, respectively. The minimum spatial frequency of the noise band was 0.5 cycles/deg for the fine-grained noise, as follows from the size of the picture, but appeared to be 2 cycles/deg for the two other types of noise. The value of σ_n was 0.45 for fine-grained noise and 0.22 for the two other types of noise. These last two values had to be multiplied with $\sqrt{(4/\pi)}$ for the calculation of Φ_n because of the circular spatial frequency limit caused by the interpolation with the Bessel function. The large bandwidth of the fine-grained noise and the medium-grained noise means that these types of noise may practically be considered as white noise. Only the coarse-grained noise is clearly nonwhite noise.

For the two-dimensional static noise used in this experiment, Eq. (2.41) of Chapter 2 gives

$$\Phi_n = \frac{\sigma_n^2}{2(u_{nmax} - u_{nmin})2(v_{nmax} - v_{nmin})} \tag{6.11}$$

so that Eq. (6.5) becomes

$$\Phi_d(u) = \frac{\sigma_n^2}{2(u_{nmax} - u_{nmin})2(v_{nmax} - v_{nmin})} \int_{u_{nmin}}^{u_{nmax}} \Psi(u_n, u) \frac{du_n}{u} \tag{6.12}$$

For two-dimensional static noise Eq. (6.1) becomes

$$m_n(u) = 2 \sqrt{\frac{\Phi_d(u)}{XY}} \tag{6.13}$$

Figs. 6.6 and 6.7 show the measurements of the contrast sensitivity function without noise and with the three types of noise for subject AVM and subject MV, respectively. The solid curves have been calculated with Eqs. (6.12) and (6.13) and Eqs. (3.26) and (6.4). For Fig. 6.6, the values of σ_0, η, and k used for the calculation were 0.6 arc min, 0.9% and 3.2, respectively, and for Fig. 6.7, these values were 0.45 arc min, 0.9% and 3.7, respectively. The general agreement between measurements and calculations is good. As expected, only the coarse-grained noise condition shows a clear effect of the bandwidth limitation. Although the maximum spatial frequency of this noise is 4.5 cycles/deg, the effect on the contrast sensitivity is still noticeable

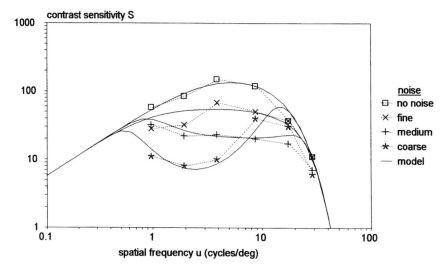

Figure 6.6: Contrast sensitivity function measured by van Meeteren & Valeton (1988) without noise and with three types of noise that differ in bandwidth. Luminance 100 cd/m². Field size 1°×1°. Binocular viewing with a natural pupil. Subject AVM. The solid curves have been calculated with Eqs. (6.12) and (6.13) and Eqs. (3.26) and (6.4).

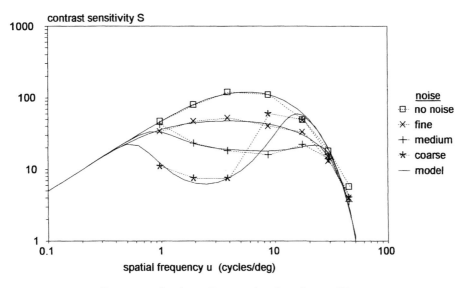

Figure 6.7: Similar as Fig. 6.6 , but for subject MV.

up to 10 cycles/deg. The given formulae for nonwhite noise are principally also valid for the more simple situation of white noise. The agreement between measurements and calculations for fine-grained noise shows that this is indeed true.

6.5 Summary and conclusions

In this chapter, a method has been given to calculate the effect of nonwhite spatial noise on contrast sensitivity. For this purpose the equations for white noise given in Chapter 2 have been generalized so that they can also be used for nonwhite noise. This was possible by the introduction of a function that describes the masking of a signal by noise components with a spatial frequency that differs from the spatial frequency of the signal. The shape of this function was derived from contrast sensitivity measurements by Stromeyer & Julesz (1972) with narrow noise bands. This function is a log normal distribution function with a 50% width of nearly two octaves. The validity of this function has been confirmed by a comparison of the model with contrast sensitivity measurements by van Meeteren & Valeton (1988) with and without white and nonwhite noise. These measurements also confirmed that formulae for nonwhite noise can principally also be used for white noise.

References

Barten, P.G.J. (1995). Simple model for spatial frequency masking and contrast discrimination. *Human Vision, Visual processing, and Digital Display VI, Proc. SPIE*, **2411**, 142-158.

Stromeyer, C.F. & Julesz, B. (1972). Spatial frequency masking in vision: critical bands and spread of masking. *Journal of the Optical Society of America*, **62**, 1221-1232.

van Meeteren, A & Valeton, J. (1988). Effects of pictorial noise interfering with visual detection. *Journal of the Optical Society of America A*, **5**, 438-444.

Chapter 7

Contrast discrimination model

7.1 Introduction

The equations for nonwhite noise given in the previous chapter can also be used for an evaluation of the masking of one signal by the presence of another signal. This application will be used for the development of a model for contrast discrimination that will be given in this chapter. For contrast discrimination, a difference has to be observed between two nearly identical sinusoidal signals that differ only in modulation. This situation is different from the situation for contrast detection where only a difference has to be observed between the presence and the absence of a signal. In contrast discrimination experiments one signal has a fixed modulation and the modulation of the other signal is varied until a just-noticeable modulation difference between the two signals is observed. The threshold of the modulation difference is the decisive quantity for contrast discrimination. The signal with the fixed modulation is called the *reference signal* and the signal with the variable modulation is called the *test signal*. The threshold of the modulation difference appears to be a function of the modulation of the reference signal. Contrast detection can be considered as a form of contrast discrimination where the reference signal has zero modulation.

To obtain a model for contrast discrimination, first an evaluation will be made of the psychometric function occurring in contrast discrimination experiments. The psychometric function, which plays an important role in contrast detection, also plays an important role in contrast discrimination. From the psychometric function of contrast discrimination experiments much information can be obtained about the fundamental aspects of contrast discrimination. This information will be used to derive an equation for the functional structure of the contrast discrimination process. This equation will further be used in the evaluation of the contrast discrimination model.

According to the theory of comparative judgment developed by Thurstone (1927a, 1927b), chance for the observation of a difference between two stimuli is a

135

function of the ratio between this difference and the uncertainty caused by the magnitude of the stimuli. For the here given model, it will be assumed that this uncertainty can be considered as a form of noise where the reference signal is considered as noise source. This noise masks the observation of the difference between test signal and reference signal. The amount of this noise can be calculated with the aid of the expressions for nonwhite noise given in the previous chapter applied on a noise source consisting of a single spatial frequency pattern. The so obtained contrast discrimination model will be compared with published data of contrast discrimination measurements.

7.2 Evaluation of the psychometric function

The psychometric function can give important information about the fundamental behavior of the visual system in contrast discrimination experiments. Fig. 7.1 shows the psychometric function for a contrast discrimination experiment by Foley & Legge (1981). In this figure the detection probability p for the modulation difference between test signal and reference signal is plotted as a function of the modulation m of the test signal. The spatial frequency of the test pattern and the reference pattern was 2 cycles/deg and the modulation of the reference pattern was 0.23%. The detection probability was calculated from the percentage of correct response of the original data in a 2AFC experiment. The data points were the result of about 200

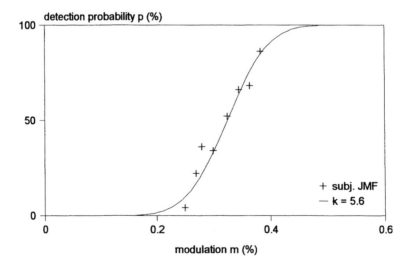

Figure 7.1: Psychometric function for a contrast discrimination experiment by Foley & Legge (1981). Spatial frequency 2 cycles/deg. Modulation of the reference signal 0.23%. The detection probability for the discrimination is plotted in the usual way as a function of the modulation m of the test signal. For the curve through the data, $k = 5.6$.

measurements made at each modulation of the reference signal (method of constant stimuli). The patterns were vertically oriented sinusoidal gratings displayed on a CRT monitor provided with P31 phosphor. The luminance was 170 cd/m² and the field size was 6°×6°. The test field was surrounded by a white surface area with a luminance that was approximately the same as the average luminance of the test field. Viewing was binocular with a natural pupil. The subject was JMF, the first author. The curve through the data has been calculated with the linear regression method mentioned in section 2.2 of Chapter 2. The experiment was made under the same conditions and with the same subject as the detection experiment of Fig. 2.2 in Chapter 2. Compared with Fig. 2.2 the slope of the curve is much steeper. The k value of this curve is 5.6, which differs considerably from 3.0 that was found for the curve in Fig. 2.2. As k must be independent of the type of experiment, the modulation m used along the horizontal axis of the figure is obviously not the right quantity that should be used for the horizontal axis of the psychometric function. In Fig. 7.2 the psychometric function is plotted for the same data with a different variable along the horizontal axis. For this variable, the modulation difference $m-m_0$ is used where m_0 is the modulation of the reference signal. Here, the slope of the curve through the data is much smaller and corresponds with a k value 1.6. This value differs also considerably from that for the detection experiment of Fig. 2.2. The slope of the function obviously depends on the quantity used along the horizontal axis. As one may assume that the k value of the visual system must have a constant value of about 3 under all conditions, a quantity was searched for the horizontal axis with which such a slope could be obtained. The result is given in Fig. 7.3 where the same data are plotted as a function of $\sqrt{(m^2 - m_0^2)}$. The curve shows here a slope corresponding with

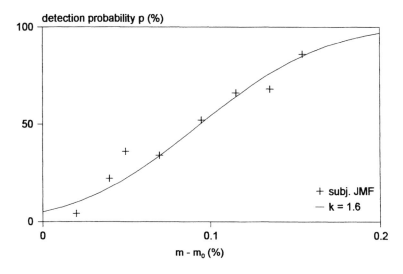

Figure 7.2: Psychometric function for the same data as Fig. 7.1, but plotted as a function of the modulation difference $m-m_0$ between test signal and reference signal. For the curve through the data, $k = 1.6$.

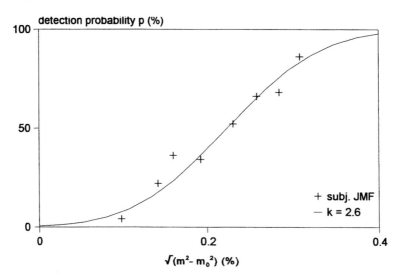

Figure 7.3: Psychometric function for the same data as Fig. 7.1, but plotted as a function of $\sqrt{(m^2 - m_0^2)}$. For the curve through the data, k = 2.6.

a k value of 2.6, which is close to the required value of 3 that was used in Fig. 2.2. It may, therefore, be assumed that the quantity $\sqrt{(m^2 - m_0^2)}$ represents the functional parameter of the contrast discrimination process. For contrast detection m_0 is zero, and the quantity becomes equal to the modulation m that was used in the previous chapters for the psychometric function of the detection process.

More support for this assumption can be obtained from measurements of the psychometric function at higher modulations of the reference signal. As mentioned in the title of their paper, the measurements by Foley and Legge used in Figs. 7.1 through 7.3 were made with a reference signal close to the detection threshold. For higher levels of the reference signal, the same phenomena can be observed, but in a more extreme way. Fig. 7.4 shows the psychometric function for a contrast discrimination experiment by Legge (1984a) at a spatial frequency of 0.5 cycles/deg with a modulation of the reference signal of 25%. This modulation is about 50 times higher than the modulation threshold for detection. The detection probability was calculated from the percentage of correct response of the original data in a 2AFC experiment. The data points were the result of about 240 measurements made at each modulation of the reference signal (method of constant stimuli). The patterns were vertically oriented sinusoidal gratings displayed on a CRT monitor provided with P31 phosphor. The luminance was 340 cd/m² and the field size was 11°×6°. Viewing was binocular with a natural pupil. The data are from one subject (DP) out of six observers, who were all in their 20's. In Fig. 7.4 the psychometric function is plotted as a function of the modulation m of the test signal. The slope of the curve corresponds with a k value 8.3. In Fig. 7.5 the same data are plotted as a function of the

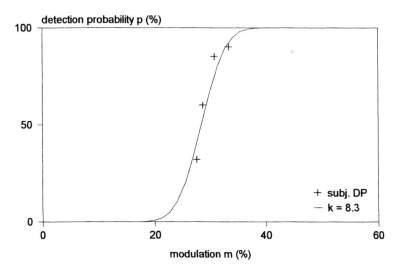

Figure 7.4: Psychometric function for a contrast discrimination experiment by Legge (1984a). Spatial frequency 0.5 cycles/deg. Modulation of the reference signal 25%. The detection probability for the discrimination is plotted in the usual way as a function of the modulation *m* of the test signal. For the curve through the data, k = 8.3.

modulation difference $m - m_0$, where m_0 is the modulation of the reference signal. The slope of the curve now corresponds with a *k* value 1.0. In Fig 7.6 the same data are plotted as a function of $\sqrt{(m^2 - m_0^2)}$. Here, the slope of the curve corresponds with a *k* value 2.8, which can be considered as the actual normal value.

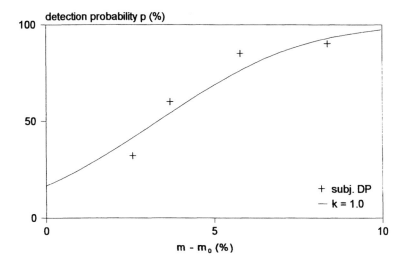

Figure 7.5: Psychometric function for the same data as Fig. 7.4, but plotted as a function of the modulation difference $m - m_0$ between test signal and reference signal. For the curve through the data, $k = 1.0$.

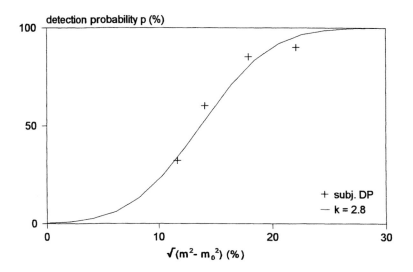

Figure 7.6: Psychometric function for the same data as Fig. 7.4, but plotted as a function of $\sqrt{(m^2 - m_0^2)}$. For the curve through the data, $k = 2.8$.

From the given data, it may be concluded that the quantity $\sqrt{(m^2 - m_0^2)}$ is in fact the functional parameter in the contrast discrimination process.

7.3 Evaluation of the contrast discrimination model

The functional parameter $\sqrt{(m^2 - m_0^2)}$ that was derived from the psychometric function in the previous section, can be written in the form $\sqrt{\{(m_0 + \Delta m)^2 - m_0^2\}}$ where m_0 is the modulation of the reference signal and Δm is the modulation difference between test signal and reference signal. At threshold, $\Delta m = \Delta m_t$, where Δm_t is the threshold of the modulation difference. This threshold depends on the modulation of the reference signal. It is now assumed here that the influence of the reference signal on the threshold can be considered as a form of noise.

At detection, the modulation threshold in the presence of external noise is given by Eq. (2.50) of Chapter 2:

$$m_t' = \sqrt{m_t^2 + k^2 m_n^2}$$

where m_t is the modulation threshold without external noise, m_t' is the increased modulation threshold with noise, and m_n is the average modulation of the noise wave components of the external noise. In analogy with this equation, we can describe the functional parameter for the threshold at contrast discrimination by

$$\sqrt{(m_0 + \Delta m_t)^2 - m_0^2} = \sqrt{m_t^2 + k^2 m_n^2} \tag{7.1}$$

where the left-hand side of the previous expression has been replaced by the functional parameter for contrast discrimination, and the modulation m_n and the right-hand side is assumed to be a function of m_0 (Barten, 1995). This equation also holds when the modulation of the reference signal m_0 is zero. Then, Δm_t is equal to m_t', if there is external noise, or Δm_t is equal to m_t, if there is no external noise. The equation can also be written in the form

$$\Delta m_t = \sqrt{m_t^2 + m_0^2 + k^2 m_n^2} - m_0 \tag{7.2}$$

The validity of this expression depends on the function used for m_n.

For this function, we assume that the reference signal can be considered as a noise source that influences the detection of the modulation difference. The reference signal consists of a single sinusoidal spatial frequency component. It is assumed here that the bandwidth of this noise is very small. For the calculation of this noise, use can be made of the equations for nonwhite noise given in the previous chapter. For m_n follows from Eq. (6.1) after omitting Y and T because of the one-dimensional static character of the noise:

$$m_n(u) = 2 \sqrt{\frac{\Phi_d(u)}{X}} \tag{7.3}$$

The function $\Phi_d(u)$ in this expression is given by Eq. (6.2). As the spatial frequency bandwidth of the noise source considered here is assumed to be small and the functions $\Psi(u_n, u)$ and $\Phi_n(u_n)$ are assumed to be constant within this frequency band (which means that $\Phi_n(u_n)$ is idealized as a rectangular distribution) Eq. (6.2) can be written in the form

$$\Phi_d(u) = \Psi(u_n, u) \Phi_n(u_n) \frac{\Delta u_n}{u} \tag{7.4}$$

where Δu_n is the bandwidth of the noise spectrum. For the situation considered here, u_n is equal to u. For this situation, it follows from Eq. (6.4) that

$$\Psi(u_n, u) = 0.747 \tag{7.5}$$

From Eq. (2.43) for the spectral density of the noise source, it follows that

$$\Phi_n = X \left(\frac{m_0}{2}\right)^2 \tag{7.6}$$

Insertion of these equations in Eq.(7.4) gives with replacement of Δu_n by Δu

$$\Phi_d(u) = 0.747 \, X \left(\frac{m_0}{2}\right)^2 \frac{\Delta u}{u} \tag{7.7}$$

Insertion of this equation in Eq. (7.3) finally gives

$$m_n(u) = \sqrt{0.747 \frac{\Delta u}{u}} \, m_0 \qquad (7.8)$$

The quantity Δu in this expression is the bandwidth of the noise spectrum. This bandwidth may be assumed to be equal to the just-noticeable spatial frequency difference between two signals with equal modulation. Campbell et al. (1970) measured this difference for a large range of spatial frequencies extending from 0.6 to 30 cycles/deg. They found that Δu is about 5 to 6% of u independent of spatial frequency and field size. Similar results were found by Burbeck & Regan (1983) and by Regan (1985) who varied also some additional conditions. This means that we may assume that $\Delta u/u \approx 0.055$. Insertion of this value in Eq. (7.8) gives

$$m_n \approx 0.2 \, m_0 \qquad (7.9)$$

and insertion of this expression in Eq. (7.2) gives

$$\Delta m_t = \sqrt{m_t^2 + (1 + 0.04 \, k^2) \, m_0^2} - m_0 \qquad (7.10)$$

Carlson & Cohen (1980) already suggested a similar equation using a parameter for the second term between the brackets that had to be found experimentally. Their model is based on the assumption that the detection of the modulation difference is determined by a constant fraction of the reference modulation. They found that the parameter for the second term between brackets is independent of luminance and field size, but varies slowly with spatial frequency. According to our model, the second term between brackets is also independent of spatial frequency, and depends only on the value of k.

If Δm_t given by Eq. (7.10) is plotted as a function of the modulation of the reference signal, a dipper-shape curve is obtained, which is well known from experiments. See the measured data shown in Figs. (7.7) through (7.9). However, at high modulations of the reference signal, the calculated curve would show a linear increase with the modulation of the reference signal, whereas the experimental data show a slight downward bend at these modulations. This bend is caused by the nonlinear behavior of the visual system at high modulation levels. Wilson (1980) proposed to take the nonlinearity into account by using a *transducer function*. As his method is rather complicated, a simpler method will be followed here. For this purpose Eq. (7.10) is first written in a form similar to Eq. (7.1):

$$\sqrt{(m_0 + \Delta m_t)^2 - m_0^2} = \sqrt{m_t^2 + 0.04 \, k^2 \, m_0^2} \qquad (7.11)$$

In this equation, the left-hand side represents the functional parameter of the contrast discrimination process. It is now assumed that the nonlinearity can be described by dividing the right-hand side of this equation by a factor $\sqrt{(1 + cm_0)}$ where c is a constant. This modifies the equation into

$$\sqrt{(m_0 + \Delta m_t)^2 - m_0^2} = \sqrt{\frac{m_t^2 + 0.04 \, k^2 \, m_0^2}{1 + cm_0}} \qquad (7.12)$$

From this equation follows that

$$\Delta m_t = \sqrt{\frac{m_t^2 + 0.04\,k^2\,m_0^2}{1 + c\,m_0} + m_0^2} - m_0 \tag{7.13}$$

It may be assumed that the constant c is determined by the ratio between the modulation m_0 of the reference signal and the average modulation of the internal noise in the visual system. This average modulation is equal to m_t/k. This means that c is some numerical factor divided by m_t/k. From a comparison with the measured data that will be given in the following sections, it appears that this factor is about 0.004. This means that

$$c \approx 0.004\;k/m_t \tag{7.14}$$

With this value of c, the final equation for the contrast discrimination threshold becomes

$$\Delta m_t = \sqrt{\frac{m_t^2 + 0.04\,k^2\,m_0^2}{1 + 0.004\,k\,m_0/m_t} + m_0^2} - m_0 \tag{7.15}$$

In the following section, the results obtained with this equation will be compared with published data of contrast discrimination measurements.

7.4 Comparison with contrast discrimination measurements

Fig. 7.7 shows contrast discrimination data measured by Nachmias & Sansbury (1974) for a vertically oriented sinusoidal grating pattern with a spatial frequency of 3 cycles/deg. The pattern was displayed on a CRT monitor provided with P31 phosphor. The field size of the stimulus was $2.2° \times 3.2°$ and was surrounded by an area with the same luminance and color as the stimulus. The luminance was not mentioned but was probably about 100 cd/m^2. The threshold was determined with a temporal 2AFC method where the threshold corresponded with 79.4% correct response. Three observers took part in the investigation. The curve through the data has been calculated with Eq. (7.15) using for k a value of 2.4 after correcting the difference between 79.4% and 75% correct response with the aid of Eq. (2.14). The calculated curve agrees well with the average results of the data.

Figs. 7.8 and 7.9 show similar measurements by Legge & Foley (1980) for a vertically oriented sinusoidal grating pattern with a spatial frequency of 2 cycles/deg. The pattern was displayed on a CRT monitor provided with P31 phosphor with a luminance of 200 cd/m^2. For the measurements given in Fig. 7.8, the field size was $0.75° \times 6°$, and for the measurements given in Fig. 7.9, the field size was $6° \times 6°$. The

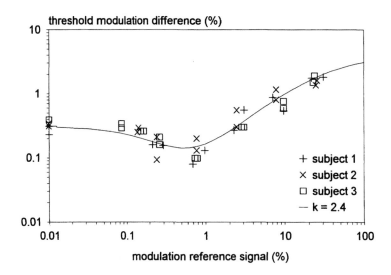

Figure 7.7: Contrast discrimination measurements by Nachmias & Sansbury (1974) for a spatial frequency of 3 cycles/deg and a field size of 2.2°×3.2°. The curve through the data has been calculated with Eq. (7.15) with k = 2.4.

test field was surrounded by a white cardboard at the same luminance, and viewed from a distance of 1.14 m. Viewing was binocular with a natural pupil. The threshold was determined with a temporal 2AFC method where the threshold corresponded with 79% correct response. Three observers took part in the investigation. The

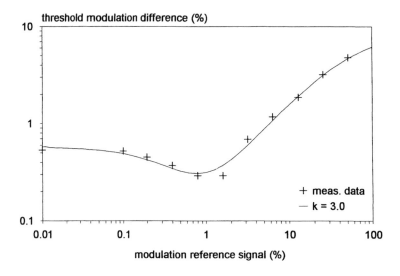

Figure 7.8: Contrast discrimination measurements by Legge & Foley (1980) for a spatial frequency of 2 cycles/deg and a field size of 0.75°×6°. Luminance 200 cd/m². Binocular viewing. The curve through the data has been calculated with Eq. (7.15) with k = 3.0.

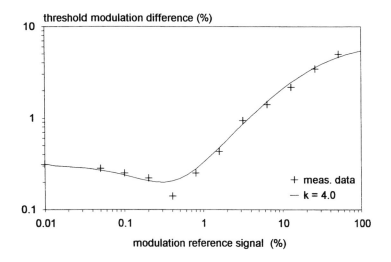

Figure 7.9: Same measurements as in Fig. 7.8 but for a field size of 6°×6°. The continuous curve has been calculated with Eq. (7.15) with k = 4.0.

reported data points were the geometric means of the measurement data. The curves through the data have been calculated with Eq. (7.15) using for *k* a value of 3.0 in Fig. 7.8 and 4.0 in Fig. 7.9 after correcting the difference between 79% and 75% correct response with the aid of Eq. (2.14). Both figures show a good agreement between measurements and calculations.

For contrast discrimination experiments where the measurements are made with monocular vision, instead of binocular vision, the same expression is valid. The only difference is that for m_t the modulation threshold for monocular viewing must be used, which is a factor $\sqrt{2}$ higher. The difference between monocular and binocular vision only occurs at low modulations of the reference signal. At higher modulations, m_t is small compared with m_0 so that the effect of m_t nearly disappears. See Eq. (7.15). This is illustrated in Fig. 7.10 where measurements by Legge (1984b) are shown made with monocular and binocular vision. The measurements were made in the same investigation as the measurements of the psychometric function shown in Figs. 7.4 through 7.6. Although the number of measurement data is small, they clearly show the expected difference with a factor $\sqrt{2}$ at a low modulation of the reference signal.

If external noise is present, Eq. (7.15) has to be extended in the following way:

$$\Delta m_t = \sqrt{\frac{m_t^2 + k^2 m_n^2 + 0.04\, k^2 m_0^2}{1 + 0.004\, k\, m_0/m_t} + m_0^2} - m_0 \qquad (7.16)$$

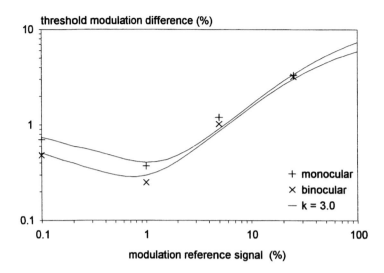

Figure 7.10: Monocular and binocular contrast discrimination measurements made by Legge (1984b) for a spatial frequency of 0.5 cycles/deg and a field size of 11°×6°. Luminance 340 cd/m². The curve through the data has been calculated with Eq. (7.15) with k = 3.0.

where m_n is the average modulation of the external noise. m_n can be calculated from the noise data with Eqs. (2.41) and (2.43) given in Chapter 2. Fig. 7.11 shows the

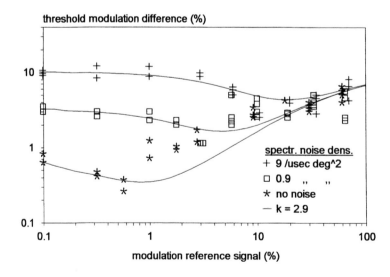

Figure 7.11: Contrast discrimination measurements by Pelli (1985) with and without two-dimensional dynamic noise for a spatial frequency 4 cycles/deg and a field size of 4°×4°. Luminance 300 cd/m². The curves through the data have been calculated with Eq.(7.16) with k = 2.9.

results of contrast discrimination measurements by Pelli (1985) with and without two-dimensional dynamic noise. He used a vertically oriented sinusoidal grating pattern with a spatial frequency of 4 cycles/deg and a luminance of 300 cd/m². The stimulus was vignetted by a Gaussian envelope to a field size of 4°×4° and an exposure time of 70 msec (1/e values). Two noise levels were used, with a spectral density of 0.9×10⁻⁶ sec deg² and 9×10⁻⁶ sec deg², respectively. However, these values had to be corrected with a factor 2³ due to a different definition of spectral noise density. The measurements at zero modulation of the reference signal have already been mentioned in Table 2.1 of Chapter 2. The threshold was determined with a temporal 2AFC method where the threshold corresponded with 82% correct response. The curves through the data have been calculated with Eq. (7.16) using for k a value of 2.9 after correcting the difference between 82% and 75% correct response with the aid of Eq. (2.14). For the noise-free situation, the agreement between measurements and calculations is less good than in the preceding figures. However, the effect of the noise is very well described.

7.5 Generalized contrast discrimination model

The curves for the threshold modulation difference Δm_t shown in the previous section depend on spatial frequency, luminance and field size, because m_t depends on these quantities. Legge (1979) already noticed that contrast discrimination curves for different spatial frequencies coincide if they are plotted in a normalized way by dividing the modulation along both axes by the modulation threshold for detection. This is confirmed by our contrast discrimination model. If we introduce in Eq. (7.15) a relative threshold for the modulation difference, defined by $\Delta m_{trel} = \Delta m_t/m_t$, and a relative modulation for the reference signal, defined by $m_{rel} = m_0/m_t$, we obtain the following generalized expression for the contrast discrimination:

$$\Delta m_{trel} = \sqrt{\frac{1 + 0.04\,k^2\,m_{rel}^2}{1 + 0.004\,k\,m_{rel}} + m_{rel}^2} - m_{rel} \qquad (7.17)$$

In this equation, Δm_{trel} does not depend on m_t, but only on m_{rel} and k. Therefore, this expression is independent from spatial frequency, luminance and field size.

Fig. 7.12 shows a plot of this generalized equation with data from measurements by Legge (1979) for four different spatial frequencies: 0.25, 1, 4, and 16 cycles/deg. The grating patterns were displayed on a CRT monitor provided with P31 phosphor with a luminance of 200 cd/m². Viewing was monocular with a natural pupil. A split-screen arrangement was used in which the reference signal and the test signal were presented on the left half of the display and a uniform field of the same luminance on the right half of the display. A vertical cardboard divider stood between the center of the display and the observer's eyes so that the left and right halves of

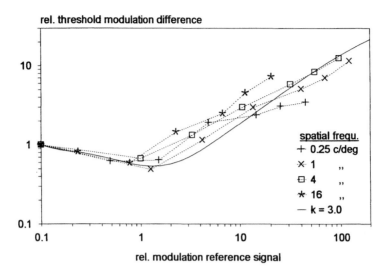

Figure 7.12: Contrast discrimination measurements by Legge (1979) for different spatial frequencies plotted with normalized coordinates. Luminance 200 cd/m². Monocular viewing with a natural pupil. The generalized curve through the data has been calculated with Eq. (7.17) with k = 3.1.

the screen were visible only to the left and right eyes, respectively. Both half-fields were 13 cm×20 cm. For the gratings of 0.25 and 1 cycles/deg, the viewing distance was 0.57 m and the half-fields subtended 13°×20°. For the gratings of 4 and 16 cycles/deg the viewing distance was 2.28 m and the half-fields subtended 3.25°×5°. The threshold was determined with a temporal 2AFC method where the threshold corresponded to 79% correct response. The observers were CF, a male in his early twenties and JG, a female of 19 years of age. The reported data points were the geometric means of 12 threshold estimates made by each of both observers. For the calculation of the generalized curve in Fig. 7.12 a *k* value of 3.0 was used after correcting the difference between 79% and 75% correct response with the aid of Eq. (2.14). Although the data points show a rather large spread, there is a reasonable general agreement between measurements and calculations.

In this investigation, Legge also made an experiment with *dichoptic viewing*. In this experiment, the reference signal was presented on the left half of the screen, whereas the test signal pattern was presented on the right half of the screen. This presentation was followed or preceded by a presentation where both halves of the screen contained only the reference signal. The observer had to identify which of the two intervals contained the test signal. Binocular fusion was aided by black fixation dots at the centers of the half-fields and by using prisms in front of the eyes.

The results of this dichoptic viewing experiment can probably be explained by modifying Eq. (7.1) that describes the functional parameter for contrast discrimination. This equation can be modified as follows:

$$\sqrt{\tfrac{1}{2}\{(m_0 + \Delta m_t)^2 - m_0^2\}} = \sqrt{m_t^2 + 2\,k^2\,m_n^2(u)} \qquad (7.18)$$

where a factor ½ has been added at the left-hand side because the test signal is presented to only one eye, so that only half of the difference is effective for the detection, and where a factor two has been added to the second term at the right-hand side because only one of the four fields contains an increase of the modulation of the reference signal, whereas otherwise one of the two fields had contained an increase of modulation. Although this last reason seems somewhat artificial, it appears to be justified by a comparison with the measurement data. Multiplication of both sides of the equation with $\sqrt{2}$ gives

$$\sqrt{(m_0 + \Delta m_t)^2 - m_0^2} = \sqrt{2\,m_t^2 + 4\,k^2\,m_n^2(u)} \qquad (7.19)$$

In this expression m_t is the detection threshold for binocular vision. If we replace this threshold for numerical reasons by the detection threshold for monocular vision, which is a factor $\sqrt{2}$ higher, we obtain

$$\sqrt{(m_0 + \Delta m_t)^2 - m_0^2} = \sqrt{m_t^2 + 4\,k^2\,m_n^2(u)} \qquad (7.20)$$

In this equation the term with $m_n(u)^2$ is multiplied with a factor 4 compared with the situation in Eq. (7.1). If we introduce this factor in Eq. (7.17), we obtain as generalized expression for the dichoptic viewing conditions of the experiment by Legge

$$\Delta m_{\text{trel}} = \sqrt{\frac{1 + 4\times0.04\,k^2\,m_{\text{rel}}^2}{1 + 0.004\,k\,m_{\text{rel}}} + m_{\text{rel}}^2} - m_{\text{rel}} \qquad (7.21)$$

Fig. 7.13 shows a plot of this equation with the results of the experiment by Legge. For k a value 3.3 was used after correcting the difference between 79% and 75% correct response with the aid of Eq. (2.14). Contrary to Fig. 7.12, the measurement data show here a small spread and a very good agreement with the calculated curve.

Fig. 7.14 shows measurements by Bradley & Ohzawa (1986) for three different spatial frequencies: 0.5 cycles/deg, 2 cycles/deg and 16 cycles/deg, plotted with normalized coordinates. The stimuli were displayed on a CRT monitor with a luminance of 250 cd/m^2 and viewed through a mask with a circular aperture of 20 cm. The mask had a uniform luminance of approximately 20 cd/m^2. The spatial frequency was varied by varying the number of cycles on the screen and by varying the viewing distance. The modulation threshold was determined with a 2AFC method where the threshold corresponded with 79% correct response. Viewing was monocular with an artificial pupil with a diameter of 2.5 mm. Two observers participated in the experiments. The data are the average results of these two observers. The solid curve has been calculated with Eq. (7.17) using a k value of 2.9 after correcting the difference

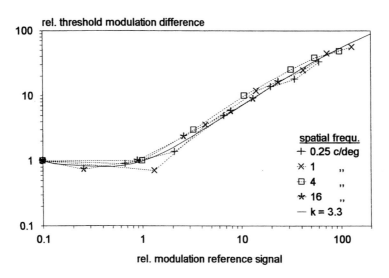

Figure 7.13: Same measurements as in Fig. 7.12 but for dichoptic viewing conditions. The generalized curve has been calculated with Eq. (7.21) with k = 3.3.

between 79% and 75% correct response with the aid of Eq. (2.14). Similarly to the measurements by Legge shown in Fig. 7.12, the measurements show some spread around the generalized curve, but they do not show a systematic dependence on spatial frequency. Although there is a reasonable general agreement between measure-

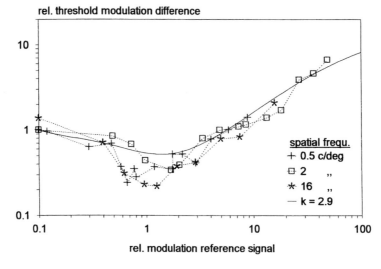

Figure 7.14: Contrast discrimination measurements made by Bradley & Ohzawa (1986) for different spatial frequencies plotted with normalized coordinates. Luminance 250 cd/m². Monocular viewing with an artificial pupil of 2.5 mm. The solid curve has been calculated with Eq.(7.17) with k = 2.9.

ments and calculations, the measurements show a tendency to a deeper minimum. However, the average deviation of the measurements from the calculated curve is opposite to the deviations shown in Fig. 7.12 for the measurements by Legge.

7.6 Summary and conclusions

In this chapter, first a principal analysis of the psychometric function has been given for the contrast discrimination between two sinusoidal luminance patterns. This analysis showed that the square root of the difference in squared modulation of test signal and reference signal is the functional parameter in the contrast discrimination process. With the aid of this parameter a model has been developed for the contrast discrimination. In this model, the effect of the reference signal on the just observable modulation difference between test signal and reference signal has been described as an effect of noise, where the reference signal is considered as noise source. The bandwidth of this noise was assumed to be equal to the just-noticeable spatial frequency difference between two signals. With the aid of the equations for nonwhite noise given in the previous chapter, the amount of noise could be expressed in the modulation of the reference signal. Furthermore, an adaptation was made for the nonlinearity of the eye at high modulation levels. In this way a contrast discrimination model was obtained that appeared to be in good agreement with published measurements. Also the effect of noise on contrast discrimination could be explained, and the effects of binocular, monocular and dichoptic viewing.

Furthermore, a generalized expression for contrast discrimination was given, which is independent of spatial frequency, luminance and field size. This expression was obtained by dividing the threshold of the modulation difference and the modulation of the reference signal by the modulation threshold for detection. The existence of such a relation was already experimentally found by Legge (1979) for spatial frequency variations, but remained until now without explanation.

References

Barten, P.G.J. (1995). Simple model for spatial frequency masking and contrast discrimination. *Human Vision, Visual processing, and Digital Display VI, Proc. SPIE*, **2411**, 142-158.

Bradley, A. & Ohzawa, I. (1986). A comparison of contrast detection and discrimination. *Vision Research*, **26**, 991-997.

Burbeck, C.A. & Regan, D. (1983). Independence of orientation and size in spatial discriminations. *Journal of the Optical Society of America*, **73**, 1691-1694.

Campbell F.W., Nachmias J., and Jukes J. (1970). Spatial-frequency discrimination in human vision. *Journal of the Optical Society of America*, **60**, 555-559.

Carlson, C.R. & Cohen, R.W. (1980). A simple psychophysical model for predicting the visibility of displayed information. *Proceedings of the SID*, **21**, 229-246.

Foley, J.M. & Legge, G.E. (1981). Contrast detection and near-threshold discrimination in human vision. *Vision Research*, **21**, 1041-1053.

Legge, G.E. (1979). Spatial frequency masking in human vision: binocular interactions. *Journal of the Optical Society of America*, **69**, 838-847.

Legge, G.E. (1984a). Binocular contrast summation-I. Detection and discrimination. *Vision Research*, **24**, 373-383.

Legge, G.E. (1984b). Binocular contrast summation-II. Quadratic summation. *Vision Research*, **24**, 385-394.

Legge, G.E. & Foley, J.M. (1980). Contrast masking in human vision. *Journal of the Optical Society of America*, **70**, 1458-1471.

Nachmias, J. & Sansbury, R.V. (1974). Grating contrast: discrimination may be better than detection. *Vision Research*, **14**, 1039-1042.

Pelli, D.G. (1985). Uncertainty explains many aspects of visual contrast detection and discrimination. *Journal of the Optical Society of America A*, **2**, 1508-1532.

Regan, D. (1985). Storage of spatial-frequency information and spatial-frequency discrimination. *Journal of the Optical Society of America A*, **2**, 619-621.

Thurstone, L.L. (1927a). Psychophysical analysis. *American Journal of Psychology*, **38**, 368-389.

Thurstone, L.L. (1927b). A law of comparative judgment. *Psychological Review*, 34, 273-286.

Wilson, H.R. (1980). A transducer function for threshold and supathreshold human vision. *Biological Cybernetics*, 38, 171-178.

Chapter 8

Image quality measure

8.1 Introduction

In this chapter, the contrast discrimination model given in the previous chapter will be used for the derivation of a measure for the perceived quality of an image. The quality of an image has always been an important aspect in the design of image forming systems. These systems can be cameras and printers for photographic systems, film projectors, display units for television, imaging systems for medical and scientific applications, etc. For the judgment of these systems a quantified measure of image quality is needed. However the design of such a measure is not easy, as the perceived quality of an image depends not only on the physical parameters of the image forming system, like resolution and contrast, but also on the impression of the image received by the eye of the observer. Therefore, the judgment by a panel of observers is often used as a quantitative measure for the obtained image quality. As this method is quite laborious and still rather subjective, investigators have searched for an objective measure for image quality in the form of a mathematical expression that contains a weighted combination of the physical parameters of the image and the psychophysical parameters of the human visual system. Such an expression is called an *image quality metric*. For such a metric, the MTF of the imaging system is generally used as physical parameter, and the contrast sensitivity function of the human eye as psychophysical parameter. The existing metrics differ from each other in the way these parameters are combined, and in the way the image quality contributions of the different spatial frequency components are taken into account. In this chapter, a survey will be given of the most important of these metrics.

Although the development of these metrics has contributed much to a better understanding of the effect of various parameters on image quality, they usually lack a good correlation with the subjectively perceived image quality. This is partly caused by the fact that in most metrics, it is assumed that the perceived image quality is linearly related with the MTF of the imaging system and therefore with the modulation of the spatial frequency components of the image. A linear relation may be valid

153

for modulations at threshold level, but the largest part of an image consists of modulations at suprathreshold level. At suprathreshold level, the nonlinear behavior of the visual system has to be taken into account. Therefore, we will first derive a model for this nonlinear behavior. This model will be based on the fact that perceived image quality appears to be linearly related with the number of discriminable modulation levels that can be perceived by the eye. For this model, use will be made of the generalized equation for contrast discrimination given in the previous section. The so obtained model appears also to give a good description of the subjectively perceived contrast as a function of modulation. The model will further be used as basis for the here given image quality metric where the nonlinear behavior of the eye is taken into account. The author proposed this metric called *square-root integral* or SQRI already earlier (Barten, 1987, 1989, 1990), but without the background that will be given here.

In this chapter, the suitability of the various metrics will be tested by a principal analysis, where the functional parameters of these metrics are compared with the requirements for a good representation of perceived image quality. Furthermore, the metrics will also be tested by a comparison of the image quality predicted for pictures with different MTFs with the subjectively perceived image quality of these pictures.

8.2 Nonlinear effect of modulation

Granger & Cupery (1972) noticed in an investigation of the image quality of photographic pictures, where several different pictures with the same content were compared, that there is a linear correlation between the perceived image quality of these pictures and the number of just-noticeable differences between these pictures. As the difference between the pictures mainly consisted of a difference in the modulations occurring in these pictures, the equations for contrast discrimination given in the previous chapter can be used to calculate the number of discriminable modulation steps for these pictures. By using Eq. (7.15) for each possible modulation, starting from zero modulation, and adding the so obtained modulation difference to the previous modulation, etc., one can calculate the number of just-noticeable differences or *jnds* as a function of the modulation. If this is done with the aid of the generalized expression given by Eq. (7.17), where the modulation is expressed in normalized units, the results are independent of spatial frequency, luminance and field size. They are shown in Fig. 8.1 for the typical k value 3. The calculation gives only data for an integer number of discrimination levels, but these data are connected in the figure by a continuous curve. Apart from the value chosen for k, this curve has general validity. Fig. 8.2 shows the same data plotted as a function of the square root of the normalized modulation. From this figure, it appears that the number of

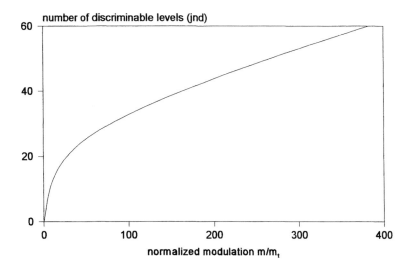

Figure 8.1: Number of discriminable modulation levels (jnds) of a sinusoidal luminance pattern as a function of the normalized modulation m/m_t. The curve has been obtained by a step by step summation of Eq. (7.17) for $k = 3$.

discriminable levels increases about linearly with the square root of the modulation. The dashed curve in this figure represents the approximation by a square-root relation. A power law for contrast discrimination was already proposed by Legge

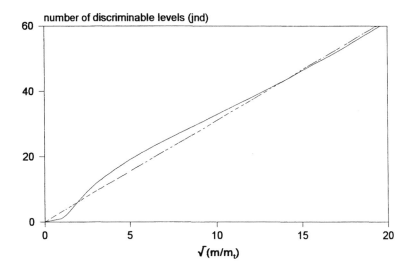

Figure 8.2: Solid curve: same as Fig. 8.1, but plotted as a function of the square root of the normalized modulation. Dashed line: approximation with a linear relation between the number of discriminable modulation levels and the square root of the normalized modulation.

(1981), based on measurements at low and medium modulation levels. From Fig. 8.2 where the modulation is extended to much higher modulation levels than in the measurements by Legge, it appears that the power law is approximately a square-root law. For normally used images, the modulation of the spatial frequency components generally extends over a large range of modulations. This means that the square-root relation forms a very good basis for an image quality metric for these images. A more precise description has no sense, as the modulations of the various spatial frequency components are arbitrarily distributed over the whole range, which cancels possible deviations. Therefore, the square root relation will be used here for the description of the nonlinear behavior of the visual system above threshold.

This relation appears also to be valid for the visually perceived contrast of sinusoidally modulated luminance patterns. Published measurements on this subject usually give the exponent of the relation between the logarithm of the visually perceived contrast and the logarithm of the modulation. However, estimating the perceived contrast is very difficult, because it is not very well defined. Furthermore, these estimations are sometimes also influenced by knowledge of the observers about the physical contrast of the object. Therefore, considerable differences of this exponent occur in these publications. The most reliable experiments in this respect have been made by Cannon (1985). He used six logarithmic spaced modulation levels between 1% and 80% modulation and four different spatial frequencies: 2, 4, 8, and 16 cycles/deg. The luminance was 100 cd/m^2 and the field size was circular with a diameter of $2°$. The average of the perceived contrast was obtained from the estimates of nine observers at two presentations. Because each subject could use a different

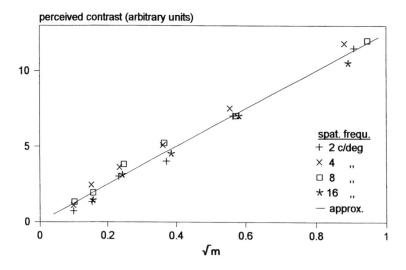

Figure 8.3: Perceived contrast as a function of the square root of the modulation for measurements by Cannon (1985) with four different spatial frequencies. The solid line shows the approximation with a linear relation between the quantities along both axes.

scale for his subjective estimate, the estimates were first normalized to a common mean of all estimates. The so obtained results were expressed in arbitrary units. They are shown in Fig. 8.3 where a linear scale is used for the perceived contrast and the modulation is plotted with a square root scale. Although the modulation threshold for the different spatial frequencies is different, they nearly coincide with a linear curve through the origin. The perceived contrast appears to be about equal at the maximum modulation and about equal at the minimum modulation, where it approaches zero. This is in good agreement with measurements by Watanabe et al. (1968) who found similar results for measurements with equally perceived contrast at different spatial frequencies. If the perceived contrast would be plotted as a function of the normalized modulation, the data would no longer approximately coincide on a common curve because of the difference in modulation threshold for the different spatial frequencies. However, it appears that they can be brought to a common curve again by dividing the perceived contrast by the square root of the modulation threshold. This is shown in Fig. 8.4. The coincidence of the data with a common linear curve for all data is even more pronounced than in Fig. 8.3.

Contrary to measurements where the object consists of a single sinusoidal luminance pattern, normal images generally consist of a combination of sinusoidal luminance patterns with different modulations and spatial frequencies. From the calculated curve in Fig. 8.2 follows that the number of discriminable modulation levels increases approximately linearly with the square root of the normalized modulation independent of spatial frequency. As Granger and Cupery found a linear

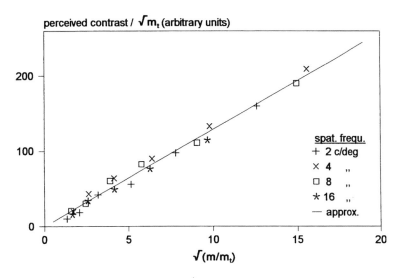

Figure 8.4: Perceived contrast divided by the square root of the modulation threshold as a function of the square root of the normalized modulation for the measurements given in Fig 8.3. The solid line shows the approximation with a linear relation between the quantities along both axes.

relation between perceived image quality and just-noticeable differences, it may be concluded that the perceived quality of an image is linearly related with the square root of the normalized modulation. The data given in Fig. 8.4 show that the perceived contrast also increases linearly with the square root of the normalized modulation. They form an extra support for the use of the square root of the normalized modulation as the functional parameter for image quality. This also means that the perceived image quality is linearly related with the square root of the modulation divided by the modulation threshold or multiplied with the contrast sensitivity. Although the contrast sensitivity of the eye is only defined at threshold level, it plays in this way still an important role at suprathreshold levels of modulation.

8.3 Image quality metrics

The modulations of the different spatial frequency components of an image are generally multiplied with the modulation transfer function or MTF of the imaging system. At low spatial frequencies the MTF is usually about 1, and at high spatial frequencies it decreases with spatial frequency. For the use of the square root of the normalized modulation in a metric for image quality, the contributions of the various spatial frequency components of an image have still to be integrated over the spatial frequency spectrum to obtain the totally perceived image quality. For the way in which these contributions have to be integrated, much can be learned from the various image quality metrics developed in the past decennia.

Image quality metrics are usually generic measures. This means that they are independent from the actual picture content. They do not contain the modulation of the different spatial frequency components of an actual image, but only the MTF by which these modulations are multiplied. This is almost remarkable, but in practice, it appears that the real amplitude of these components does not play an important role in the judgment of image quality. The amplitude of the components is obviously taken into account as a generic quantity in the distribution of the different spatial frequency components over the spatial frequency spectrum.

As sharpness has always been one of the most important aspects of image quality, early attempts for image quality metrics were concentrated on resolution. This resolution was usually expressed in a number of lines that could just be distinguished. As this number describes only the maximum resolution of an image, it can only be used as a comparative measure for types of images that are similar in other aspects. A more precise definition of resolution became possible with the use of the MTF in the early fifties, which was mainly due to the pioneering work of Schade (1951-1955). With the aid of the MTF, a more exact description could be given of the reduction of sharpness at high spatial frequencies.

8.3.1 Modulation transfer area (MTFA)

However, for a good description of image quality, the contrast sensitivity of the eye also has to be taken into account. One of the first attempts to take the contrast sensitivity of the eye into account was made by Charman & Olin (1965). They introduced an image quality metric that they originally called *threshold quality factor* or TQF, but was later promoted by Snyder (1973) under the name *modulation transfer area* or MTFA. In this metric the surface area between the MTF and the modulation threshold as a function of spatial frequency is used as a measure of image quality. The MTFA is given by

$$A = \int_0^{u_{max}} \{M(u) - m_t(u)\} \, du \tag{8.1}$$

where $M(u)$ is the MTF of the imaging system, $m_t(u)$ is the modulation threshold of the eye, and u_{max} is given by the condition

$$M(u_{max}) = m_t(u_{max}) \tag{8.2}$$

The unit of this metric is spatial frequency. The metric is based on the idea that a modulation below the threshold cannot contribute to visual perception. Although the MTFA is widely used as a measure for image quality and is still used as official standard in the USA, it lacks a good correlation with subjectively perceived image quality. This is partly caused by the fact that it is linearly related with the modulation, instead of with the square root of it, and partly caused by the fact that the modulation threshold is subtracted from the MTF, instead of dividing the MTF by it.

8.3.2 Integrated contrast sensitivity (ICS)

To overcome the last of these two disadvantages, van Meeteren (1973) proposed a metric called *integrated contrast sensitivity* or ICS. In this metric the modulation threshold is not subtracted from the MTF, but the MTF is multiplied by the contrast sensitivity of the eye, which means that it is divided by the modulation threshold. The ICS is given by

$$I = \int_0^{\infty} M(u)S(u) \, du \tag{8.3}$$

where $S(u)$ is the contrast sensitivity function. Similarly as with the MTFA, the image quality is expressed in spatial frequency units. Although the ICS is already an improvement over the MTFA, it still depends linearly on modulation.

8.3.3 Subjective quality factor (SQF)

In the MTFA and the ICS, the contribution of the different spatial frequency components to the total image quality is taken into account by linear integration over spatial frequency. Granger & Cupery (1972) found that logarithmic integration shows a better correlation with perceived image quality. They introduced a metric called *subjective quality factor* or SQF. In this metric a logarithmic integration over the spatial frequency is used. Similarly as with the ICS, the MTF is multiplied by the contrast sensitivity function, but the contrast sensitivity function is simplified to a rectangular function with a value 1 between 3 and 12 cycles/deg and a value zero outside this range. The SQF is given by

$$Q = K \int_{u_1}^{u_2} M(u) \, \mathrm{d}(\log u) \tag{8.4}$$

where $u_1 = 3$ cycles/deg, $u_2 = 12$ cycles/deg, and K is a normalizing constant equal to $1/\log 4$ to normalize the SQF to a value 1 for the ideal situation that $M(u) = 1$ over the total integration range. In this way, the image quality is expressed in dimensionless units. Although the logarithmic integration is an improvement over a linear integration, the dependence on modulation is still linear. Furthermore, the representation of the contrast sensitivity function by a rectangular function forms an oversimplification of the actual situation, as was found later by Higgins (1977) and will be shown in Fig. 8.8 of section 8.5.

8.3.4 Discriminable difference diagram (DDD)

As has already been mentioned in the section 8.2, Granger and Cupery also found in their investigation, that there is a linear correlation between subjective image quality and just-noticeable differences. Carlson & Cohen (1980) used this principle for the development of an image quality metric based on the number of discriminable modulation differences in different spatial frequency areas. Instead of a logarithmic integration over the spatial frequencies, they applied a logarithmic summation by using logarithmic spaced spatial frequency channels with a width of a factor two. The discriminable modulation levels of each channel were indicated in a *discriminable-difference diagram* or DDD. The total number of just-noticeable differences under the MTF of the imaging system gives the resulting image quality. They assumed that a difference of one unit of this number is equal to a difference of one jnd in image quality. The results of their image quality metric are, therefore, expressed in units of just-noticeable differences. Although the use of the jnd as unit for image quality is an improvement over the use of other units, the method is not very practical. For every condition of luminance and field size, a different diagram has to be used. The constants required for the calculation of these diagrams have to be obtained from measurements at these conditions. Although the dependence on modulation is nonlinear, the DDD does not always show a good correlation with perceived image quality. Furthermore, the just-noticeable difference derived from the sum of the discriminable differences in the different spatial frequency channels appears not to correspond with the just-noticeable difference between two images.

8.3.5 Square-root integral (SQRI)

As improvement over the DDD and the other existing metrics, the author proposed some years ago a new metric called *square-root integral* or SQRI (Barten 1987, 1989, 1990). In this metric, the nonlinear behavior of the eye is taken into account by making use of the assumption given in section 8.2 that the perceived image quality is linearly related with the number of just-noticeable modulation differences and that this number is linearly related with the square root of the modulation divided by the threshold modulation. Furthermore, use is made of the fact that the modulation of the different spatial frequencies of an image can be represented by the multiplication factor formed by the MTF of the imaging system, as is done in other metrics, and a logarithmic integration over the spatial frequencies is used, as was done by Granger and Cupery in their SQF metric (See section 8.3.3). The image quality given by this metric is expressed in units of just-noticeable differences or jnds, as was done by Carlson and Cohen in their DDD metric (See section 8.3.4). The SQRI is given by

$$J = \frac{1}{\ln 2} \int_{u_{min}}^{u_{max}} \sqrt{\frac{M(u)}{m_t(u)}} \, \frac{du}{u} \qquad (8.5)$$

where du/u stands for the logarithmic integration over spatial frequency. The constant in front of the integral has been chosen so that one unit of the SQRI corresponds indeed with one just-noticeable difference in image quality. To obtain this, the constant has been determined from a comparison with measurement data published by Carlson & Cohen (1980). The constant appeared to be about 1.4 or about $1/\ln2$. See Barten (1987). Although the choice of $1/\ln2$ for this constant has no physical meaning, it may be useful for an interpretation of the results. It says that the SQRI increases with one unit when the integrand increases with one unit in a spatial frequency band with a width of a factor two. This is the width of the spatial frequency channels used by Carlson and Cohen in their DDD metric. The jnd, which was already introduced by Carlson and Cohen, is a very important unit for image quality, because it is a basic unit of perception. It has further the advantage over other possible units that it enables a good interpretation and comparison of the results. A difference of one jnd corresponds with 75% correct response in a 2AFC experiment.

In the SQRI, integration limits are used to restrict the integration to spatial frequency areas that can contribute to the image quality. For television images, for instance, the available spatial frequency area is limited by the bandwidth of the television system. Higher spatial frequencies cannot contribute to the perceived image quality. For the same reason also a lower limit has been introduced. This limit is caused by the finite spatial dimensions of the image. When the frequency spectrum is written as a Fourier sum, the lowest term of this sum has a frequency $1/X$, where X is the size of the image. With an assumed bandwidth $1/X$ for each term in the continuous representation used for the integral, the lowest frequency is $0.5/X$. This means that for u_{min}, half the inverse of the image size has to be used.

8.4 Two-dimensional aspects

A remarkable aspect of the image quality metrics discussed in the previous section is that they are expressed by a one-dimensional equation. One would have expected a two-dimensional equation because of the two-dimensional character of an image. The one-dimensional form of these metrics is, however, only apparent. They can also be written in a two-dimensional form. If the one-dimensional expression is represented by an integral of the form

$$\int F(u)\,du \tag{8.6}$$

this integral can be written as

$$\frac{1}{2\pi}\int F(u)\,\frac{2\pi u\,du}{u} \tag{8.7}$$

One can now convert this in a two-dimensional expression by replacing $2\pi u du$ by $du dv$ and $1/u$ by $1/\sqrt{(u^2+v^2)}$. This transforms the one-dimensional integral in

$$\frac{1}{2\pi}\int\int F(u,v)\,\frac{du\,dv}{\sqrt{u^2+v^2}} \tag{8.8}$$

where the function $F(u)$ has been replaced by the function $F(u,v)$ that may be different for different directions. From Eq. (8.6) also another two-dimensional expression can be derived by using polar coordinates. This gives

$$\frac{1}{2\pi}\int_0^{2\pi}\left\{\int F(u,\vartheta)\,du\right\}d\vartheta \tag{8.9}$$

This expression is equivalent to an averaging of the one-dimensional integral over different directions. In practice, four directions are already sufficient: $0°$, $90°$, $+45°$, and $-45°$. Images are usually isotropic and the contrast sensitivity of the eye is also almost isotropic, so that the one-dimensional form of a metric is usually already sufficient.

8.5 Functional analysis of image quality metrics

By making use of the linear relation between perceived image quality and just-noticeable differences, different image quality metrics can be analyzed with respect to their effect on image quality. For this analysis, image quality measurements will be used where the image quality is varied in steps of one just-noticeable difference.

For this analysis, the equation of an image quality metric is written in the following general form:

$$J = \int j(u) \, d(\ln u) \tag{8.10}$$

where J stands for the total image quality expressed in jnds and $j(u)$ is a distribution function that gives the image quality contribution per logarithmic spatial frequency interval $d(\ln u)$. For practical reasons a logarithmic spatial frequency interval has been chosen, although a linear interval du or a quadratic interval du^2 would also have been possible. From the equations for the various metrics given in the previous sections, the following expressions can be derived for the distribution function $j(u)$:

for the MTFA

$$j(u) \propto \{M(u) - 1/S(u)\} \, u \tag{8.11}$$

for the ICS

$$j(u) \propto M(u) \, S(u) \, u \tag{8.12}$$

for the SQF

$$j(u) \propto \frac{1}{\ln 4} M(u) \quad \text{for } 3 \text{ c/deg} < u < 12 \text{ c/deg}, \quad \text{else } j(u) = 0 \tag{8.13}$$

and for the SQRI

$$j(u) = \frac{1}{\ln 2} \sqrt{M(u) \, S(u)} \tag{8.14}$$

With exception of the SQRI, the functions $j(u)$ still have to be multiplied with a certain factor to express J in jnds.

In Fig. 8.5 a schematic drawing is given for the MTFs of two focus conditions that differ one jnd in image quality. The spatial frequencies where the MTFs have decreased with 50% are u_1 and u_2. If the focus condition of an image is varied in steps of one just-noticeable difference, the function $j(u)$ can be determined from measurements of u_1 and u_2 and be compared with the above given expressions. If the MTFs are approximated by rectangular functions with a width given by u_1 and u_2, the function $j(u)$ can be derived from these frequencies with the aid of the following expression:

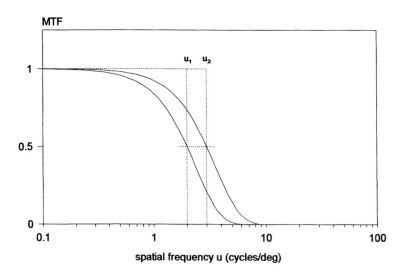

Figure 8.5: Schematic drawing of two MTFs that give a difference of one jnd in image quality. These MTFs can be idealized by rectangular functions with a width determined by the spatial frequencies u_1 and u_2 where the MTFs have decreased with 50%.

$$j(u) = \frac{\mathrm{d}J}{\mathrm{d}(\ln u)} \approx \frac{\Delta J}{\Delta(\ln u)} = \frac{1}{\ln u_\theta - \ln u_\Theta} = \frac{1}{\ln(u_\theta/u_\Theta)} \tag{8.15}$$

For the spatial frequency u in $j(u)$ the geometric mean of u_1 and u_2 may be used. The experiment has to be repeated for several values of u_1 to get the image quality contribution for a range of values of the spatial frequency spectrum.

Fig. 8.6 shows the results obtained from an experiment with this type of measurement made by Carlson & Cohen (1980). They varied the focus condition of projected slides by a just-noticeable change in sharpness over a wide range of focus conditions, and measured the spatial frequency where the MTFs had decreased to 50%. The MTFs had a Gaussian shape. The average luminance was 111 cd/m^2 (35 mL) and the field size was 25.6°×25.6°. The images consisted of a crowd scene, a portrait of a woman's face in color, and a monochrome version of the same picture. The first picture contained many small details, whereas the two other pictures had large smooth transitions. Although the data points in Fig. 8.6 show a considerable scattering, their general behavior is consistent and shows no systematic difference between the different picture types. This confirms the assumption that the perceived image quality can be described by a generic quantity, which does not depend on the actual picture content. The solid curve through the data points gives the SQRI prediction calculated with the aid of Eq. (8.14). This curve agrees very well with the average of the data points.

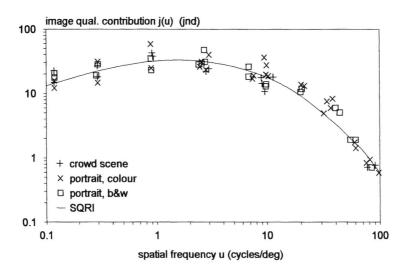

Figure 8.6: Image quality contribution of different spatial frequency areas calculated with Eq. (8.15) for measurements by Carlson & Cohen (1980). The different symbols represent different types of images. The data show no systematic difference for the different types of images. The solid curve through the data points gives the SQRI prediction calculated with the aid of Eq. (8.14).

The same type of measurements was made by Carlson & Cohen (1978) with projected slides of an airplane cockpit display. The average luminance was 6.4 cd/m^2

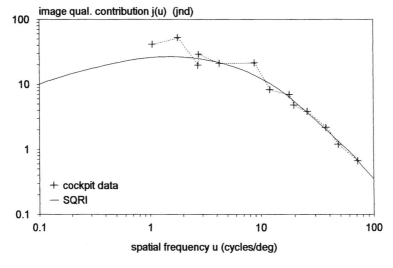

Figure 8.7: Same as Fig . 8.6, but for an image with artificial data of an airplane cockpit display. The solid curve through the data points gives the SQRI prediction calculated with the aid of Eq. (8.14). The similarity with Fig. 8.6 shows that there is no effect of image content on the perceived image quality.

(2 mL) and the field size was $7.2° \times 7.2°$. Fig. 8.7 shows the results obtained from these measurements. Although the picture is the image of an artificial object, it shows about the same behavior as the natural images in Fig. 8.5. The solid curve through the data points gives the SQRI prediction calculated with the aid of Eq. (8.14). This curve too agrees very well with the data points.

In Fig. 8.8 the same data are shown as in Fig. 8.6, but plotted with a linear vertical scale, instead of a logarithmic scale and using only the geometric means of the data, as the spread of the data points would have increased considerably by plotting them on a linear scale. Besides the image quality contribution predicted by the SQRI, the predictions by the MTFA, the ICS and the SQF metrics are also shown. For these metrics arbitrary factors were used to convert the units to jnds. The use of a linear vertical scale in this figure means that the surface area under the curves corresponds with the total image quality. Therefore, the figure gives a good impression of the relative importance of the different spatial frequency areas. From the figure, it can be seen that the MTFA gives an underestimation of the image quality contribution of low spatial frequencies and a strong overestimation of the image quality contribution of high spatial frequencies. The ICS gives a much better estimate at high frequencies than the MTFA, but still gives an underestimation of the image quality contribution of low spatial frequencies. The SQF uses only a small part of the total spatial frequency range and gives, therefore, an underestimation of low spatial frequency areas and high spatial frequency areas. The SQRI clearly shows the best agreement

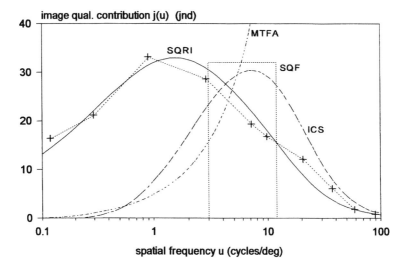

Figure 8.8: Same data as in Fig. 8.4, but plotted with a linear vertical scale and using only the geometric mean of the data. In addition to the image quality contribution predicted by the SQRI, also the predictions by the MTFA, ICS, and SQF metrics are shown. These contributions were multiplied by arbitrary factors to convert the units to jnds. The area below the curves represents the total image quality.

with the results obtained from these measurements.

Unfortunately, there are only a few measurements available where the image quality is changed in steps of one just-noticeable difference. In experiments by other investigators, the focus condition of the image was changed by an electronic blurring process. The difference in focus was measured by determining the change of the sigma of the line-spread function used in the blurring process. For a Gaussian line-spread function with a standard deviation σ, the MTF is given by

$$M(u) = e^{-2\pi^2\sigma^2 u^2} \tag{8.16}$$

From this equation follows for the spatial frequency where the MTF is 50%

$$u_{0.5} = \sqrt{\frac{\ln 2}{2}}\,\frac{1}{\pi\sigma} \tag{8.17}$$

With this equation the values of u_1 and u_2 used in Eq. (8.15) can be obtained. Replacing these values by the corresponding sigma values σ_1 and σ_2 gives

$$j(u) = \frac{1}{\ln(\sigma_1/\sigma_2)} \tag{8.18}$$

The value of u in $j(u)$ can be calculated with an equation similar to Eq. (8.17) by using the geometric mean of σ_1 and σ_2.

Fig. 8.9 shows the so obtained data of the image quality contribution for measurements by Baldwin (1940). He measured the just-noticeable focus difference of projected movie pictures at different focus conditions. The average luminance was

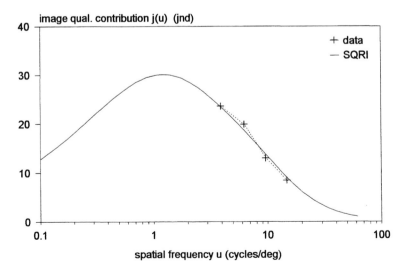

Figure 8.9: Image quality contribution of different spatial frequency areas calculated with Eq. (8.18) for measurements of defocused movie pictures by Baldwin (1940). The solid curve gives the SQRI prediction calculated with the aid of Eq. (8.14).

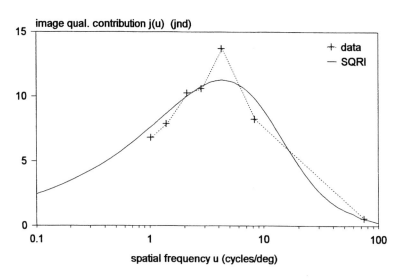

Figure 8.10: Image quality contribution of different spatial frequency areas calculated with Eq. (8.18) for measurements of a blurred edge transition by Watt & Morgan (1983). The solid curve gives the SQRI prediction calculated with the aid of Eq. (8.14).

7 cd/m^2 and the field size was $14.5° \times 13.7°$. The just-noticeable differences were obtained from a measurement of the psychometric function. The solid curve in the figure represents the SQRI prediction calculated with Eq. (8.14). This curve agrees very well with the data. The use of the psychometric function for the measurements has probably caused that the deviations are very small. Predictions by the other image quality metrics have been omitted, because they show the same kind of deviations as were shown in Fig. 8.8.

Fig. 8.10 shows similar results for measurements made by Watt & Morgan (1983). They used band-shaped images consisting of a single edge transition displayed on a CRT monitor. The average luminance was 292 cd/m^2, and the field size was $3° \times 0.2°$. The images were electronically blurred in the direction of the transition. The just-noticeable differences were determined with a 2AFC method. The data points are the average for two subjects. In this figure too, the solid curve through the data points has been calculated only for the SQRI. The different shape of the curve compared with the shapes of the curves in Figs. 8.5 and 8.6 is caused by the small field size in vertical direction. With a few exceptions, the position of the data points agrees well with the SQRI prediction.

In the given experiments only the spatial frequency was changed. At these frequencies, it was assumed that the MTF dropped from a value 1 to a value zero for the hypothetical situation of a rectangular MTF. The question remains how the image quality would vary if the spatial frequency is constant and only the modulation or the

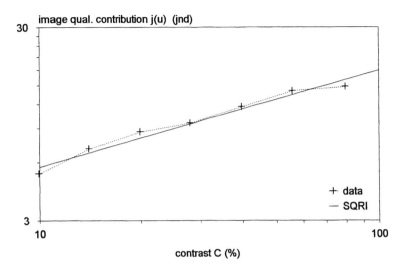

Figure 8.11: Image quality contribution as a function of contrast calculated with Eq. (8.18) for measurements of a blurred edge transition by Watt & Morgan (1983).The data show a square-root dependence on contrast. The solid line through the data gives the SQRI prediction calculated with Eq. (8.14) after multiplying M with the contrast C.

MTF is changed. In the MTFA, ICS and SQF metrics a linear relation with the modulation is assumed, whereas in the SQRI a square-root relation is assumed.

Watt and Morgan made an experiment in a second part of their investigation, where they changed the contrast at a fixed focus condition and measured the just-noticeable change in focus as a function of contrast. In the same way as for their other experiment, the image quality contribution $j(u)$ could be derived from these measurements, but now as a function of contrast at a fixed spatial frequency. This spatial frequency could be calculated from the average sigma of the blur and appeared to be 4.5 cycles/deg. The results for the average data of two observers are shown in Fig. 8.11. The data points show a square-root dependence on contrast. The solid line through the data points represents the prediction by the SQRI calculated with Eq. (8.14) after multiplying M with the contrast factor C. The SQRI prediction shows a very good agreement with the data. The square-root dependence on contrast of the data points over a large range of contrast values confirms the validity of the square-root approximation. The MTFA, ICS and SQF predictions are not given in the figure. They would have shown a linear dependence on contrast.

8.6 Effect of differently shaped MTFs

In the previous section, different image quality metrics have been compared with respect to their effect on image quality by making use of the linear relation between perceived image quality and the number of just-noticeable differences. From the given analysis, it might be clear that most image quality metrics will not show a good agreement with actually perceived image quality. One may wonder why these metrics sometimes still show a good correlation with perceived image quality. The reason is that MTFs often have a similar, for instance, Gaussian shape, so that the comparison between different images is limited to a small type of variations. Under these conditions, even the worst image quality metric can show a good correlation with perceived image quality. If the spread of the measurement data is very large, even a fundamentally better metric can hardly improve the correlation. Correlation coefficients say often more about the spread in the measurement data than about the quality of the metric.

However, a larger difference between different metrics occurs when images are compared that are displayed with differently shaped MTFs. Experiments with such MTFs have been made by Higgins (1977). He measured the subjective image quality of four photographic images that were each displayed with 22 different MTFs. Some of these MTFs had a quite irregular shape, like the examples shown in Fig. 8.12. The images had a size of 10×10 cm and were viewed at a distance of 0.38 m, corresponding with a field size of $15°$. The reported data are the averages of the judgments by 20 observers. Figs. 8.13 through 8.16 show comparisons of the observed image quality with the image quality calculated with the MTFA, ICS, SQF, and SQRI, respectively. For all four metrics a linear regression has been made between measurements and calculations. The solid lines in the figures are the regression lines. The R^2 value of the

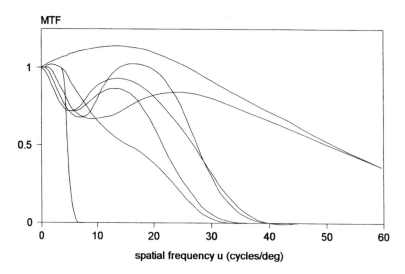

Figure 8.12: Seven of the 22 different MTFs used by Higgins (1977) in his investigation for a comparison of image quality metrics. From the examples shown here, some have a very irregular shape.

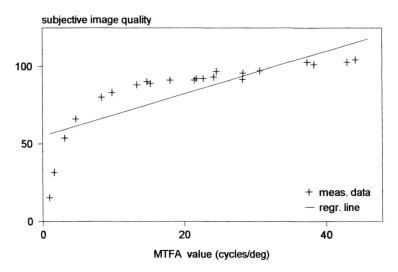

Figure 8.13: Measured subjective image quality as a function of MTFA value for photographs reproduced with 22 different MTFs from an investigation by Higgins (1977). The solid line is the linear regression line. Correlation between measurements and calculations 63.0%.

correlation between measurements and calculations was 63.0% for the MTFA, 93.3% for the ICS, 89.3% for the SQF and 99.5% for the SQRI. This order of succession roughly corresponds with the difference in image quality contribution shown in Fig.

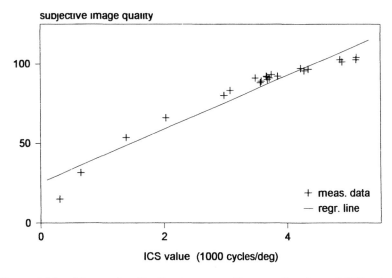

Figure 8.14: Measured subjective image quality as a function of ICS value for photographs reproduced with 22 different MTFs from an investigation by Higgins (1977). The solid line is the linear regression line. Correlation between measurements and calculations 93.3%.

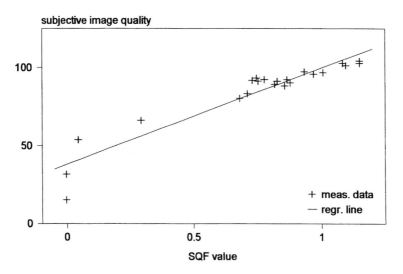

Figure 8.15: Measured subjective image quality as a function of SQF value for photographs reproduced with 22 different MTFs from an investigation by Higgins (1977). The solid line is the linear regression line. Correlation between measurements and calculations 89.3%.

8.8. Fig 8.16 shows that the relation between measurements and calculations for the SQRI is strictly linear over the very large variation of the image quality used in this experiment. The fit with the data can hardly be improved. The remaining spread is

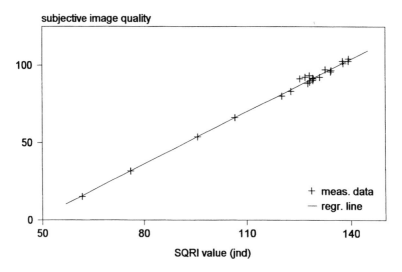

Figure 8.16: Measured subjective image quality as a function of SQRI value for photographs reproduced with 22 different MTFs from an investigation by Higgins (1977). The solid line is the regression line. Correlation between measurements and calculations 99.5%.

probably caused by a spread of the measurement data.

8.6 Summary and conclusions

In this chapter a model has been given for the nonlinear behavior of the eye at suprathreshold levels of modulation. This model is important for a good description of perceived image quality, as the components of an image largely consist of modulations at suprathreshold level. According to this model, the perceived quality of an image is linearly related with the square root of the normalized modulation. This is based on the fact that image quality is linearly related with the number of discriminable levels, and that from the expressions for contrast discrimination given in the previous chapter follows that this number is linearly related with the square root of the normalized modulation. Furthermore, it has been shown that also the perceived contrast of sinusoidal luminance patterns is linearly proportional with the square root of the normalized modulation.

The perceived quality of an image is usually described with the aid of an image quality metric. In this chapter a survey has been given of various image quality metrics developed during the past decennia. Also an image quality metric has been given that is based on the square root of the normalized modulation, according to the model given in the first part of this chapter. This metric is called SQRI or square-root integral.

In a principal test, the functional parameters of these metrics have been compared with the requirements for a good description of perceived image quality. For this analysis use has been made of experiments where the image quality was varied in steps of one just-noticeable difference. From the analysis, it appeared that only the SQRI metric shows a good agreement with the required behavior.

The metrics have also been tested by a comparison of predicted and measured image quality for photographic pictures displayed with different MTFs. In this test, the SQRI showed the best correlation with the measured data. The relation was strictly linear over a large range of image qualities.

References

Baldwin, M.W. (1940). The subjective sharpness of simulated television images. *Bell System Technical Journal*, **19**, 563-587.

Barten, P.G.J. (1987). The SQRI method: a new method for the evaluation of visible resolution on a display. *Proceedings of the SID*, **28**, 253-262.

Barten, P.G.J. (1989). The square root integral (SQRI): a new metric to describe the effect of various display parameters on perceived image quality. *Human Vision, Visual Processing, and Digital Display I, Proc. SPIE*, **1077**, 73-82.

Barten, P.G.J. (1990). Evaluation of subjective image quality with the square-root integral method. *Journal of the Optical Society of America A*, **7**, 2024-2031.

Cannon, M.W. (1985). Perceived contrast in the fovea and the periphery. *Journal of the Optical Society of America A*, **2**, 1760-1768.

Carlson, C.R. & Cohen, R.W. (1972). Visibility of displayed information: image descriptors for displays. Report ONR-CR213-120-4F, Office of Naval Research, Arlington, VA.

Carlson, C.R. & Cohen, R.W. (1980). A simple psychophysical model for predicting the visibility of displayed information. *Proceedings of the SID*, **21**, 229-246.

Charman, W.N. & Olin, A. (1965). Image quality criteria for aerial camera systems. *Photographic Science and Engineering*, **9**, 385-397.

Granger, E.M. & Cupery, K.N. (1972). An optical merit function (SQF), which correlates with subjective image judgments. *Photographic Science and Engineering*, **16**, 221-230.

Higgins, G.C. (1977). Image quality criteria. *Journal of Applied Photographic Engineering*, **3**, 53-60.

Legge, G.E. (1981) A power law for contrast discrimination. *Vision Research*, **21**, 457-467.

Schade, O. (1951-1955). Image gradation, graininess, and sharpness in television and motion picture systems. *Journal of the SMPTE*, **56**, 137-171, **58**, 181-222, **61**, 97-164, and **64**, 593-617.

Snyder, H.L. (1973). Image quality and observer performance. *Perception of displayed information, L.M. Biberman, Editor*, Chapter 3, Plenum Press, New York and London.

van Meeteren, A. (1973). Visual aspects of image intensification. Ph. D. Thesis, University of Utrecht, Utrecht, The Netherlands.

Watanabe, A., Mori, T., Nagata S., and Hiwatashi, K. (1968) Spatial sine-wave responses of the human visual system. *Vision Research*, **9**, 1245-1263.

Watt, R.J. & Morgan, M.J. (1983) The recognition and representation of edge blur: evidence for spatial primitives in human vision. *Vision Research*, **23**, 1464-1477.

Chapter 9

Effect of various parameters on image quality

9.1 Introduction

In the previous chapter, various metrics were analyzed for their suitability for the description of image quality. In these metrics, use was made of the MTF of the imaging system and the contrast sensitivity of the eye. From these metrics, the SQRI appeared to have the best relation with perceived image quality. The SQRI was originally developed for the evaluation of the effect of resolution on perceived image quality (Barten, 1987), but later, it appeared that it could also be used for the description of the effect of many other parameters on image quality (Barten, 1989a, 1990a). In this chapter, the effect of a number of parameters on image quality will be treated. They will be analyzed with the aid of the SQRI and will largely be compared with published measurements. For the contrast sensitivity of the eye, Eq. (3.26) in Chapter 3 will be used with the there given typical values of the constants.

Beyond the MTF of the imaging system, the contrast sensitivity of the eye plays an important role in the perception of image quality. Generally, some parameters influence the MTF, other parameters influence contrast sensitivity, and still other parameters influence both MTF and contrast sensitivity. In the SQRI the following parameters can be taken into account: **Resolution** is taken into account in the MTF. **Addressability** or addressable resolution is taken into account in the upper limit of the integration. Addressability is determined by the number of lines or the number of pixels with which the image is reproduced. **Luminance,** defined here as the average luminance of the image, influences contrast sensitivity and is therefore taken into account in the contrast sensitivity function. **Contrast,** defined by the size of the luminance variations divided by the average luminance, is taken into account in the MTF, because contrast variations cause a multiplication of the MTF with a certain factor. **Gamma** (see section 9.7) is also taken into account in the MTF, as it influences image quality in a similar way. **Viewing distance** is both taken into account in the contrast sensitivity function, because of its effect on field size, and in the MTF, because of its effect on angular spatial frequency. **Image size** at a constant

175

viewing distance has the same effects as viewing distance and can therefore be treated in the same way. **Noise** in an image increases the modulation threshold and is therefore taken into account in the contrast sensitivity function. **Luminance quantization** used in digital images causes small local deviations from the original luminance and can therefore be considered as a form of noise. **Pixel geometry** used in matrix displays influences the addressable resolution and can therefore be taken into account in the upper limit of the integration.

All these effects will shortly be treated in this chapter and the predicted results will as much as possible be compared with published measurements. The image quality measurement are usually obtained from the judgments by a panel of observers that look at the images under carefully standardized viewing conditions. For the judgment, a *category scaling* method is generally used. In some cases, the subjects have to express their opinion in a scale number, and in other cases a standard series of expressions is used, like "bad," "poor," "fair," "good," and "excellent." These expressions are then later transformed into a number. The results of the category scaling are often modified to transform them into a linear perceptual scale by using a technique based on Thurstone's law of comparative judgment (Togerson, 1958). By this technique, the scale is linearized to obtain a scale where the distances are perceived as equal differences. This scale is comparable with the jnd scale used in the SQRI. In some measurements, two or more parameters are varied simultaneously. It appears that subjects still can give a consistent judgement of the total image quality. This means that the visual system is able to weight the different parameters in a fixed relative proportion to each other. For these measurements the validity of a metric is of special importance, as it has to weight the different parameters in the same relative proportion as the visual system.

For a comparison of the measured data with the calculated data, a linear regression will be made, and the correlation coefficient R^2 will be calculated. This coefficient can be used for an interpretation of the results, as $1 - R^2$ is the part of the variance which has to be attributed to the sum of the inaccuracy of the measurements and a systematic deviation between theory and practice. The constants of the regression equation will be used to convert the calculated SQRI values in the image quality units that were used in the experiment. This enables an easy comparison between measurements and calculations. Both scale units will be used in the graphical representation of the results.

9.2 Resolution and image size

The effect of resolution on image quality was measured by Westerink & Roufs (1989) together with the effect of image size in an investigation where both parame-

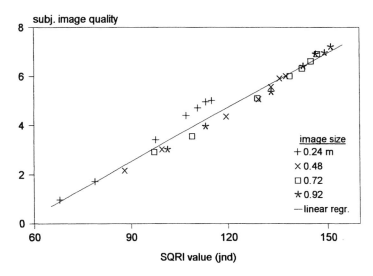

Figure 9.1: Linear regression between subjective image quality and SQRI value for measurements by Westerink & Roufs (1989) with color slides projected with different resolutions and sizes. Average luminance 30 cd/m². Viewing distance 2.9 m. The SQRI values have been calculated with Eq. (8.5). The correlation between measurements and calculations is 96.3%.

ters were varied simultaneously. For this investigation, projected color slides were used with different resolutions and sizes. The images were square pictures of five

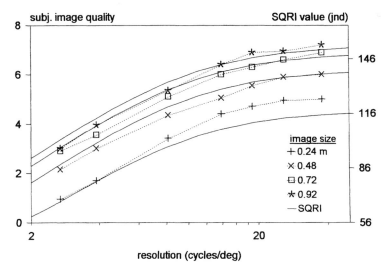

Figure 9.2: Measurement data of Fig. 9.1 plotted as a function of resolution with the image size as parameter. The resolution is expressed in the spatial frequency where the MTF has decreased with 50%. The solid curves have been calculated with the SQRI with the aid of Eq. (8.5).

different pictorial scenes. They were projected with an average luminance of 30 cd/m², and viewed from a distance of 2.9 m. The size of the image was varied from 0.24 m to 0.92 m, corresponding with a field size ranging from 4.7° to 18°. The variation was obtained by using copies of the slides in four different sizes. The resolution was varied by defocusing the projector with the aid of a stepper motor. The MTF of the projector had a Gaussian shape. The resolution was expressed in the spatial frequency where the MTF has decreased to 50%. The image quality was determined by using a 10-point numerical category scale. Twenty subjects took part in the experiment. The average rating was used for the results. Fig. 9.1 shows a linear regression between the measured and calculated data. The correlation is 96.3%. In Fig. 9.2 the measurements and calculations are plotted as a function of resolution with the image size as parameter. In this figure, the resolution scale is expressed in the spatial frequency where the MTF has decreased to 50%. The curves for the different image sizes are approximately parallel and show a saturation at a resolution of about 20 cycles/deg.

9.3 Luminance and image size

The effect of luminance on image quality was measured by Van der Zee & Boesten (1980) of the same laboratory, together with the effect of image size, in an investigation where both parameters were varied simultaneously. For this investigation, also projected color-slides were used of which the luminance was varied with the aid of

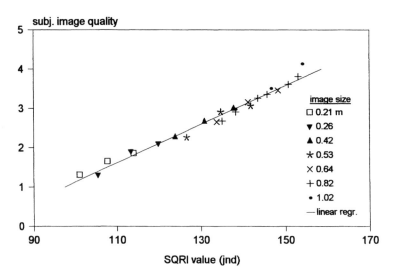

Figure 9.3: Linear regression between subjective image quality and SQRI value for measurements by van der Zee & Boesten (1980) with color slides projected with different luminance and sizes. Viewing distance 2.9 m. The SQRI values have been calculated with Eq. (8.5). The correlation between measurements and calculations is 97.7%.

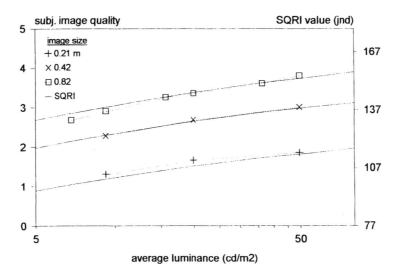

Figure 9.4: Measurement data of Fig. 9.3 plotted as a function luminance with image size as parameter. For clearness the results for only three image sizes are shown. The solid curves have been calculated with the SQRI with the aid of Eq. (8.5).

two projectors. For this investigation, the same equipment with largely the same slides was used as later was used in the investigation by Westerink and Roufs treated in the previous section. Two images of the same scene - differing only in luminance and size - were projected simultaneously with the aid of the two projectors. The viewing distance was 2.9 m. The projected image size was varied in seven steps from 0.21 m to 1.02 m, corresponding with a field size ranging from 4.2° to 20°. The luminance was varied with the aid of neutral density filters and was determined by measuring the open gate luminance. The average luminance may be assumed to be one tenth of the open gate luminance, similarly as in the experiments by Westerink and Roufs with the same slides. The perceived image quality was determined by using a 5-point category rating scale. Twenty-nine subjects took part in the experiments. The quality ratings were averaged over the 29 subjects. Fig. 9.3 shows a linear regression between measured and calculated data. The correlation is 97.7%. In Fig. 9.4 the measurements and calculations are plotted as a function of luminance with the image size as parameter. For clearness only the results for three image sizes are shown. Curves for the other sizes are similar. The agreement between measurements and calculations is very good, certainly if it is considered that two parameters were varied simultaneously. The calculated curves are approximately parallel to each other. They show a saturation at a level of about 100 cd/m^2.

9.4 Anisotropic resolution

Although the resolution of displayed images is usually equal in all directions, this need not always be so. If the resolution of an image is different in different directions, the two-dimensional description given in section 8.4 of the previous chapter has to be used for the analysis of the effect of resolution on image quality. This will be illustrated with measurements by Nijenhuis (1993) which were made in the same laboratory as the experiments mentioned in the two previous sections. He measured the perceived image quality of a blurred image on a CRT monitor at several combinations of horizontal and vertical blur. The blur was electronically generated with the aid of a Gaussian spread function, and was characterized by the standard deviation σ of this function. The image was an artificial picture consisting of an evenly lit square area of 10 cm × 10 cm with a luminance of 45 cd/m² on a background of 28 cm × 28 cm with a luminance of 10 cd/m². The average luminance of this image was 15.7 cd/m² and the viewing distance was 4 m, which corresponds with a field size of 4°×4°. For the evaluation of the results, the SQRI was calculated for four directions: the horizontal direction, the vertical direction, and the two diagonal directions. The average for these four directions was used for the calculation of the final results. For the diagonal direction, it was assumed that the sigma of the blur can be derived from the sigma of the blur in horizontal and vertical direction by using the following

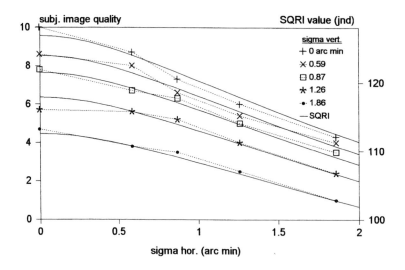

Figure 9.5: Subjective image quality of a blurred image measured by Nijenhuis (1993) as a function of the horizontal blur with the vertical blur as parameter. The blur is expressed in the standard deviation sigma of the spread function used in the blurring process. Average luminance 15.7 cd/m². Field size 4°×4°. The solid curves have been calculated with the SQRI averaged over four directions, using Eqs. (8.5), (8.9), and (9.1). The correlation between measurements and calculations is 98.7%.

equation:

$$\sigma_{dia} = \sqrt{\frac{1}{2}\sigma_{hor}^2 + \frac{1}{2}\sigma_{ver}^2} \qquad (9.1)$$

Measurement data and calculation results are shown in Fig. 9.5 as a function of the horizontal blur with the vertical blur as parameter. The agreement between measurements and calculations is very good. The correlation amounts to 98.7%. From the figure, it can be seen that the curves of a constant vertical blur are not parallel but converge slightly at increasing horizontal blur. If the SQRI had been calculated by taking only the average over the horizontal and vertical directions, the curves had been parallel and the correlation between measurements and calculations would have been worse.

9.5 Viewing distance, display size, and number of scan lines

The perceived quality of an image is often influenced by the viewing distance. For the European 625 lines PAL (Phase Alternation Line) television system, the optimum viewing distance is assumed to be six times the picture height. For the American 525 lines NTSC (National Television System Committee) television system, this distance is assumed to be eight times the picture height. These values are based on practical experience and are confirmed by recommendations of the CCIR (Comité Consultatif International de Radiocommunication). At a given image size, a variation of the viewing distance means a variation of the field size. Furthermore, the angular resolution of the image is changed. Both factors have an opposite effect on image quality. This causes an optimum distance where the image quality is maximum. The same holds for a chance in display size at a fixed viewing distance. This means that there is also an optimum display size at a given viewing distance. The optimum viewing distance or the optimum image size of a television system can be calculated in the following way (Barten, 1990b):

For a television system, the resolution of the image is limited by the bandwidth of the system. This limitation can be taken into account in the SQRI in the upper limit of the integration. For the maximum spatial frequency of a television system holds

$$u_{max} = \frac{K N_v}{2h} \qquad (9.2)$$

where N_v is the visible number of scan lines, h is the height of the image expressed in angular size for the eye, and K is the *Kell factor*. This is a factor by which the vertical resolution has to be reduced to avoid interference effects between the scan lines of the system and high spatial frequency components in vertical direction of the original

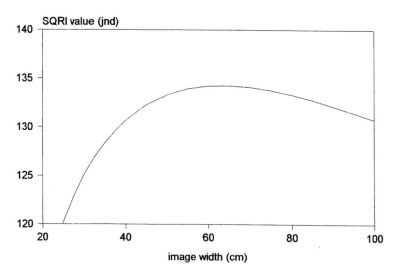

Figure 9.6: Perceived image quality as a function of image width, calculated with the SQRI for a PAL television picture viewed at a distance of 2.5 m. The calculation has been made with the aid of Eqs. (8.5) and (9.2). A maximum occurs at a width of about 60 cm. At this width, the viewing distance is about six times the image height.

image. For K usually a value 0.7 is used. The bandwidth of the television system has generally been chosen such that the maximum spatial frequency in horizontal direction is equal to that in vertical direction. This means that the resolution is equal in both directions. The visible number of scan lines can be calculated from the total number of scan lines by subtracting the losses by vertical retracing (usually 8%) and the losses by vertical overscan (usually 6%). For the NTSC and PAL systems furthermore a resolution loss of 29% and 26%, respectively, has to be taken into account because of the presence of the color sub-carrier. For a CRT display, a Gaussian MTF may be assumed that is mainly influenced by the size of the electron spot. The width of this spot at 5% of its maximum may be assumed to be about 0.4% of the picture width.

Under these conditions, the SQRI has been calculated for a PAL television picture with an average luminance of 100 cd/m^2, and a viewing distance of 2.5 m. This calculation was made for various image sizes. The results are shown in Fig. 9.6 as a function of the image width. From the figure, it can be seen that the calculated curve shows indeed a maximum. At a lower size, the image quality decreases because of a decrease in field size, and at a higher size, the image quality decreases because of a decrease in angular resolution. The maximum occurs at an image width of about 60 cm, which corresponds with an image height of about 45 cm. At this size, the viewing distance is about six times the image height, which is considered as the optimum viewing distance for the PAL system.

9.6 Contrast

The contrast of an image can be defined as the difference between the maximum luminance and the minimum luminance divided by the sum of them. According to this definition the maximum contrast is one. Stray light or reflected ambient light generally causes an equal increase of maximum and minimum luminance, whereas the difference remains the same. This gives a reduction of contrast. For the sinusoidal components of an image, this means that the average luminance is increased, whereas the amplitude of the luminance variation has remained the same. This means that the modulation of each spatial frequency component of an image is multiplied with a factor C given by

$$C = \frac{\bar{L}}{\bar{L} + \Delta L} = \frac{1}{1 + \frac{\Delta L}{\bar{L}}} \tag{9.3}$$

where $\bar{L}$ is the average luminance without added light and ΔL is the amount of added light (Barten, 1989b). This factor is the same for all spatial frequencies. The reduction of the modulation with this factor also means that the MTF of the imaging system has to be multiplied by this factor. As the SQRI is proportional to the square root of the MTF, the SQRI has to be multiplied by the square root of C. So that

$$J' = \sqrt{C}\, J \tag{9.4}$$

where J is the SQRI value without contrast loss and J' is the SQRI value with contrast loss. A contrast loss of, for instance, 6% corresponding with $C = 0.94$ would lead to a reduction of the SQRI value with 3%, which is a loss of image quality of about three jnds. The loss of image quality is often perceived as a loss of sharpness. This is caused by the multiplication of the MTF by the contrast factor, which appears to have the same effect on perceived sharpness as a reduction of the MTF.

For most types of displays, the contrast loss by ambient light can be reduced by the use of dark tinted screen glass. This goes, however, at the cost of the average luminance of the image. From one hand, the image quality increases by the increase of contrast, but from the other hand, the image quality decreases by the decrease of the luminance. There is, however, an optimum glass transmission where the image quality reaches a maximum. This optimum can be calculated with the SQRI by using the above given expressions (Barten, 1991c).

For a reflective image, like a photograph, the contrast factor C is generally equal to the reflectivity of the material used for the photographic print. Here, the amplitude of the luminance variations is multiplied by the reflectivity of the material, whereas the average luminance is usually determined by the luminance of the surrounding area. An investigation where contrast loss in photographic images appeared to play an important role was made by Feng and Östberg (Feng, Östberg,

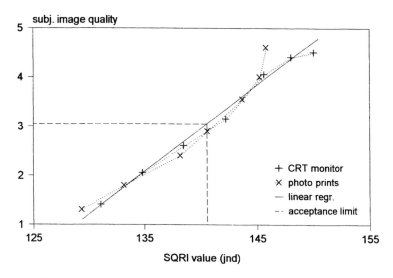

Figure 9.7: Linear regression between subjective image quality and SQRI value for measurements by Feng and Østberg (1990, 1991) with text images displayed on a CRT and on photo prints. Luminance 100 cd/m². Field size 23.7°×31.3°. The dashed line represents the acceptance limit. The SQRI has been calculated with the aid of Eq. (8.5). In the calculation, it has been taken into account that the photo prints had a lower contrast. The correlation between measurements and calculations is 96.5% for the combined results of the two types of images.

and Lindström, 1990, Östberg & Feng, 1991). They measured the perceived image quality of text images that were displayed with different focus conditions on a CRT together with images on photographic material that were copies of the same images. The luminance was 100 cd/m² and the viewing distance was 0.5 m, corresponding with a field size of 23.7°×31.3°. The photo prints were viewed under the same conditions as the CRT images, but the reflectivity of the material for the photographic prints was not 100% but 86% (Barten, 1992). Fig. 9.7 shows a linear regression between the subjective image quality and the calculated SQRI value for the results of both types of images. In the calculations, the contrast reduction of the photo prints was taken into account. The correlation between measurements and the calculations for the combined results of the two types of images is 96.5%. The dashed line represents the acceptance limit that the authors found in their investigation. In the figure no systematic difference can be seen between the two types of images. Östberg & Feng (1991) found, however, unexplained systematic differences between the curves for the two types of images. However, they did not take the contrast loss of the photo prints into account and used, furthermore, the MTFA as a measure for image quality. If the contrast loss was not taken into account in the calculations for Fig. 9.7, the curve for the photo prints would have been shifted to the right over about nine jnds.

9.7 Gamma

The relation between input and output luminance of an imaging system is not always linear. If it is nonlinear, it can often be described by an exponential relation given by

$$L' = \text{const. } L^{\gamma} \qquad (9.5)$$

where L' and L are the output luminance and the input luminance, respectively, and γ is the exponent of the exponential relation. The relation can also be written in the form

$$\log L' = \text{const. } + \gamma \, \log L \qquad (9.6)$$

From this equation follows, that if L' is plotted as a function of L on a double logarithmic scale, a straight line is obtained with a slope equal to gamma.

For a gamma different from one, the modulation transfer from input to output cannot be described by an MTF, as the MTF concept is based on Fourier analysis that can only be applied to linear systems. With an exponential relation between the input and the output luminance, the output modulation depends on the luminance in the concerning part of the image. If gamma is higher than one, local luminance differences increase in light parts of the image but decrease in dark parts. If gamma is lower than one, the inverse takes place. However, we found that, if these effects are averaged over all luminance levels occurring in an image, the modulation increases approximately linearly with gamma (Barten, 1996). This means that the average effect of gamma can be described by a multiplication of the MTF by gamma. As the SQRI is proportional with the square root of the MTF, the SQRI value and consequently also the image quality will increase linearly with the square root of gamma.

In practice, it appears that this is indeed so at low gamma values. However, at gamma values above an optimum of about 1.2, the perceived image quality decreases again at a further increase of gamma. This decrease can be explained by a loss of discriminable details that disappear in the dark and light areas of the image. From measurements by van Hateren & van der Schaaf (1996, Fig. 2), it appears that the luminance distribution of natural scenes can roughly be described by a rectangular function of the logarithm of the luminance. Fig. 9.8 shows a schematic representation of this distribution. From Eq. (9.6) follows that the width of this distribution on logarithmic scale increases proportionally with gamma. If gamma becomes higher than the optimum value, a part of the image is lost. From the probability density distribution shown in Fig. 9.8, it can be seen that the remaining part of the image is equal to γ_0/γ, if γ_0 is the optimum value of gamma. From experiments, it appears that for gamma values higher than γ_0, the perceived image quality decreases linearly with this remaining part, besides an increase with the square root of gamma due to the further increase of the modulation in this remaining part. The variation of the perceived image quality at a variation of gamma can, therefore, be described by the

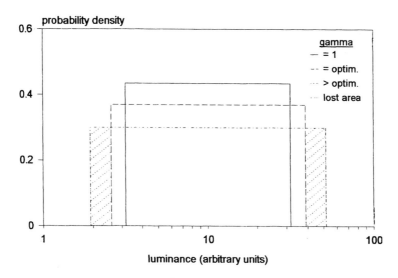

Figure 9.8: Schematic representation of the probability density distribution of the luminance in natural images and the change of this distribution at an increase of gamma. If gamma becomes higher than the optimum value, image details are lost in dark and light parts of the image.

following equations:

$$J' = \gamma^{0.5} J \quad \text{for } \gamma < \gamma_0 \tag{9.7}$$

and

$$J' = \gamma^{0.5} \left(\frac{\gamma_0}{\gamma} \right) J = \left(\frac{\gamma_0}{\gamma^{0.5}} \right) J \quad \text{for } \gamma \geq \gamma_0 \tag{9.8}$$

where J is the SQRI value for $\gamma = 1$, and J' is the SQRI value for other values of γ. According to this model that we already published earlier (Barten, 1996), the image quality increases with the square root of gamma up to an optimum gamma value, and decreases again at higher gamma values inversely with the square root of gamma.

Fig. 9.9 shows an evaluation of measurements by Roufs et al. (1994) with the aid of Eqs. (9.7) and (9.8). They measured the subjectively perceived image quality as a function of gamma over a large range of gamma values using colored pictures of three natural scenes: a town hall, a terrace scene, and a woman's portrait. The original objects were provided with luminance bars at the edge of the scenes for a calibration of gamma in the final images. The images were displayed on a video monitor of which the gamma could electronically be varied, while keeping the average luminance constant. The luminance was about 25 cd/m^2. The size of the pictures was 28 cm × 28 cm and they were viewed at a distance of 2.1 m, corresponding with a field size of 6.8°×6.8°. The ratings were made with a 10-point numerical scale. The measurement data are the averages of the ratings by four observers. The optimum gamma was 1.35.

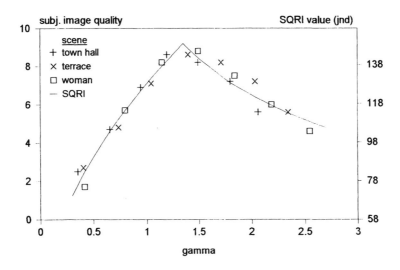

Figure 9.9: Subjective image quality as a function of gamma measured by Roufs et al. (1994) for three natural images displayed on a CRT monitor. Luminance 25 cd/m². Field size 6.8° × 6.8°. The curve through the data has been calculated with the SQRI with the aid of Eq.(8.5), using Eqs. (9.7) and (9.8) for the effect of gamma. Optimum gamma 1.35. The correlation between measurements and calculations 95.3%.

The correlation between the measurements and calculations is 95.3%.

Fig. 9.10 shows a similar evaluation of measurements by Shimodaira et al. (1995). They measured the perceived image quality of five digital standard test charts of the ITEJ (Institute of Television Engineers of Japan) for ten different gamma values. The images were displayed on an 8.6-inch TFT-LCD (thin film transistor liquid crystal display) and viewed at a distance of six times the screen height, which corresponds with a field size of 12.7°×9.5°. The average luminance was about 14 cd/m². The image quality judgments were made with a 5-point rating scale. The given data are the averages of the ratings by 15 observers. The optimum gamma was 1.13. The optimum gamma for each of the separate pictures differed slightly from each other. This causes a rounding of the maximum for the average of the pictures. The correlation between measurements and calculations is 98.2%.

Fig. 9.11 shows similar results for measurements by Mitsubayashi et al. (1996) with largely the same team of authors as for the previous investigation. They measured the perceived image quality of five digital standard test charts of the ITEJ (Institute of Television Engineers of Japan) for ten different gamma values. The images were now displayed on a 14-inch CRT monitor used at three different luminance levels: 6, 14, and 40 cd/m². As the results for these luminance levels did not differ much from each other, only the data for 14 cd/m² were used for the evaluation given here. The images were viewed at a distance of six times the screen

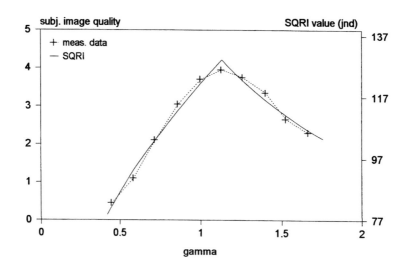

Figure 9.10: Subjective image quality as a function of gamma measured by Shimodaira et al. (1995) for the average of five test images displayed on a TFT-LCD. Luminance 14 cd/m². Field size 12.7°×9.5°. The curve through the data has been calculated with the SQRI with the aid of Eq. (8.5), using Eqs. (9.7) and (9.8) for the effect of gamma. Optimum gamma 1.13. The correlation between measurements and calculations is 98.2%.

height, which corresponds with a field size of 12.7°×9.5°. The image quality judg-

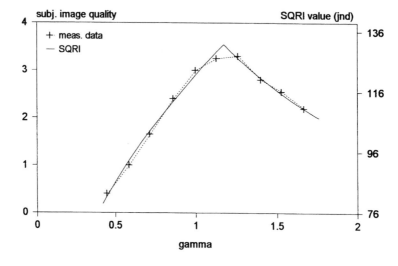

Figure 9.11: Subjective image quality as a function of gamma measured by Mitsubayashi et al. (1996) for the average of five test images displayed on a CRT monitor. Luminance 14 cd/m². Field size 12.7°×9.5°. The curve through the data has been calculated with the SQRI with the aid of Eq. (8.5), using Eqs. (9.7) and (9.8) for the effect of gamma. Optimum gamma 1.17. The correlation between measurements and calculations is 99.3%.

ments were made with a 5-point rating scale. The data were averaged over observers and images. The optimum gamma value was 1.17. The correlation between measurements and calculations is 99.3%. From Figs. 9.9 through 9.11, it can be seen that the model used for the evaluation of the effect of gamma on perceived image quality agrees very well with the measurements.

9.8 Noise

For the effect of noise on image quality, we assume that it can be described by the increase of the modulation threshold due to the noise (Barten, 1991a). This increase can be calculated with the aid of Eqs. (2.50) and (2.43) given in Chapter 2. This will be illustrated with the following examples of different types of experiments.

Besides the measurements with different MTFs given in section 8.6 of the previous chapter, Higgins (1977) also measured the image quality of photographic pictures with different amounts of noise. These measurements were made in combination with three different resolutions by using three different MTFs in the photographic reproduction process before adding the noise. The further conditions were the same as for the investigation described in section 8.6 of the previous chapter. By the

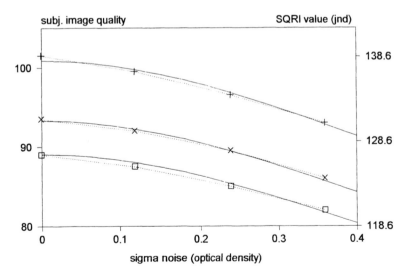

Figure 9.12: Subjective image quality measured by Higgins (1977) as a function of the sigma of the noise for photographic images with three different resolution levels. The sigma of the noise is expressed in units of optical density. The solid curves have been calculated with the SQRI with the aid of Eq. (8.5), using Eqs. (2.50), (2.43) and (2.41) for the dependence on noise. The correlation between measurements and calculations is 99.7%.

simultaneous variation of noise and resolution, a good impression can be obtained of the relative weight of these parameters on image quality. Measurements and calculations are shown in Fig. 9.12. In this experiment, the sigma of the noise was expressed in units of optical density. The correlation between measurements and calculations is 99.7%. The good agreement between measurements and calculations for this experiment confirms the assumption that the effect of noise on image quality can simply be described by an increase of the modulation threshold.

Kayargadde (1995) made a study of different aspects of noise. He did not measure the perceived image quality, but the perceived noisiness. However, the opposite of perceived noisiness can be considered as a measure for perceived image quality. The reported noisiness will, therefore, be compared here with the calculated SQRI value after reversing its sign. Furthermore, the original data of the standard deviation of the noise will be expressed in relative units with respect to the average luminance. This quantity will shortly be called sigma noise.

In one experiment noise was added to a pure white field displayed on a CRT monitor. The images had a size of 0.17 m × 0.17 m and were viewed from a distance of 1.4 m corresponding with a field size of 7°×7°. Two different luminance levels were used: 20 cd/m² and 30 cd/m². Subjects were asked to judge the noisiness of the picture. The data are the average results from eight subjects. Fig. 9.13 shows the so obtained data with reversed sign as a function of the sigma of the noise for the

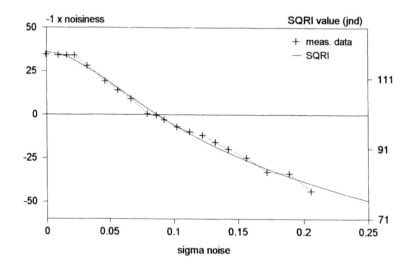

Figure 9.13: Noisiness as a function of the sigma of the noise measured by Kayargadde (1995) for a white field on a CRT monitor. The noisiness has been plotted with a reversed sign. Luminance 30 cd/m². Field size 7°×7°. The solid curve has been calculated with the SQRI with the aid of Eq. (8.5), using Eqs. (2.50), (2.43) and (2.41) for the dependence on noise. The correlation between measurements and calculations is 99.4%.

luminance level of 30 cd/m². The results for 20 cd/m² are similar. The solid curve through the data has been calculated with the SQRI. The correlation between measurements and calculations is 99.4%. The agreement between the measurements and the calculations is somewhat remarkable, as the image just consisted of an evenly lit white field. In this field no pictorial information is present that could be masked by the noise. The observed noisiness is obviously perceived as a degradation in quality of the flat image.

It is well known that noise in a dark image is more annoying than noise in a light image. One would therefore expect that noise in a dark part of an image is also more annoying than noise in a light part. Within an image, the relative standard deviation of the noise is the same for the dark and light parts, because the average luminance of the total image is the same for both areas. This would give the same effect on image quality for the dark and the light parts. The question rises if this is the right way to treat the effect of noise on image quality. Kayargadde analyzed this aspect of noise by measuring the perceived noisiness as a function of the local luminance in a part of the image, while keeping the average luminance in the image constant. The image consisted of nine square blocks with five different luminance levels ranging from 1.9 to 29.5 cd/m². The average luminance was 12.8 cd/m². The total size of the image was 0.2 m × 0.2 m and the viewing distance was 1.4 m,

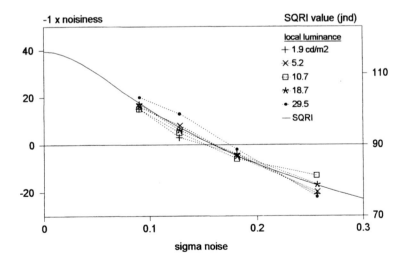

Figure 9.14: Noisiness as a function of the sigma of the noise measured by Kayargadde (1995) for the central part of an image consisting of blocks with different luminance. The local luminance of the central block was varied. The noisiness has been plotted with a reversed sign. Average luminance of the total image 12.8 cd/m². Field size 8.2°×8.2°. The solid curve has been calculated with the SQRI with the aid of Eq. (8.5), using Eqs. (2.50), (2.43) and (2.41) for the dependence on noise. The correlation between measurements and calculations is 95.9%. The measurement data show no systematic dependence on local luminance.

corresponding with a field size of 8.2°×8.2°. The noise was added only in the central block of the image. Subjects were asked to judge the noisiness in this part. The luminance of this block was changed to each of the five luminance levels used by exchanging the luminance of this block with one other block of the image. The measurement data are the average results from eight subjects. Fig. 9.14 shows the so obtained data as a function of the relative standard deviation of the noise with the local luminance as parameter. The solid curve through the data has been calculated with the SQRI. For the calculation of the relative standard deviation of the noise used in the SQRI, the average luminance of the total image was used. The correlation between measurements and calculations is 95.9%. The measured data show no systematic difference between the different local luminance levels, although these differ up to a factor 15 from each other. Only the results for the highest luminance level seem to differ systematically from the other luminance levels, but this difference inverses at the highest noise level. From the results of this experiment, it may be concluded that only the average luminance of an image plays a role for the effect of noise on image quality, even if the different parts of an image differ considerably in luminance.

The luminance deviations caused by noise usually show a Gaussian distribution. However, the distribution need not always be Gaussian. Kayargadde investigated the effect of different distributions of the noise over the luminance. For this

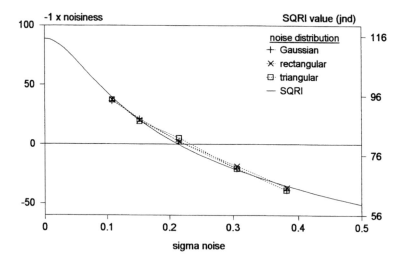

Figure 9.15: Noisiness as a function of the sigma of the noise measured by Kayargadde (1995) for a white field with three different noise distributions. The noisiness has been plotted with a reversed sign. Luminance 20 cd/m². Field size 7°×7°. The solid curve has been calculated with the SQRI with the aid of Eq. (8.5), using Eqs. (2.50), (2.43) and (2.41) for the dependence on noise. The correlation between measurements and calculations is 99.4%. There is no systematic difference between the different noise distributions.

purpose he used besides a Gaussian distribution, a rectangular distribution and a triangular distribution. In all these situations, the amount of noise was characterized by the standard deviation of the luminance deviation. For the SQRI, the sigma of the noise is also determined by this standard deviation, as follows from Eq. (2.31) given in Chapter 2. The noise was added to a white field with a luminance of 20 cd/m². The size of the image was 0.17 m × 0.17 m and the viewing distance was 1.4 m, corresponding with a field size of 7°×7°. Subjects were asked to judge the noisiness of the image. The measurement data are the average results from eight subjects. Fig. 9.15 shows the so obtained data for the three different noise distributions as a function of the sigma of the noise. The solid curve through the data has been calculated with the SQRI. The correlation between measurements and calculations is 99.4%. The measured data show no systematic difference for the different distributions. This means that for the calculation of the effects of noise, only the standard deviation of the luminance deviations is important, regardless of the type of distribution of these deviations over the luminance.

Kayargadde also measured the effect of noise on natural images: a terrace scene and a woman's portrait. The size of the image was 0.24 m × 0.24 m and the viewing distance was 1.4 m, corresponding with a field size of 9.8°×9.8°. Subjects were asked to judge the noisiness of the image. The measured data are the average results from eight subjects. Fig. 9.16 shows the so obtained data as a function of the sigma of the noise. The solid curve has been calculated with the SQRI. As expected, the results for

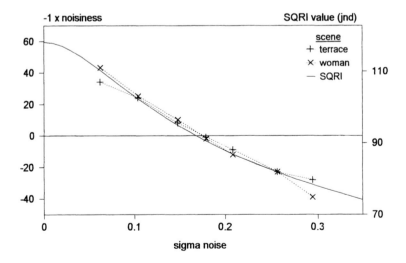

Figure 9.16: Noisiness as a function of the sigma of the noise measured by Kayargadde (1995) for two natural scenes. The noisiness has been plotted with a reversed sign. Luminance 9 cd/m². Field size 9.8°×9.8°. The solid curve has been calculated with the SQRI with the aid of Eq. (8.5), using Eqs. (2.50), (2.43) and (2.41) for the dependence on noise. The correlation between measurements and calculations is 97.9%.

both pictures do not show a systematic difference. The correlation between measurements and calculations is 97.9% for the combined results of both pictures.

9.9 Pixel density and luminance quantization

Images are nowadays usually stored and transferred in digital form. For the conversion in digital format, the images are spatially sampled and the luminance information is quantified. Also at the reproduction of an image on a matrix display, like a LCD (liquid crystal display), ELD (electro luminescence display) or PDP (plasma display panel), the information is spatially sampled. Spatial sampling sets a limit to the obtainable resolution and luminance quantization causes local deviations from the original continuous luminance variation.

In the SQRI, the resolution limit caused by spatial sampling can be taken into account in the upper limit of the integration. The resolution limit is given by the Nyquist frequency of the pixel structure

$$u_{max} = u_N = \frac{1}{2p} \qquad (9.9)$$

where u_N is the Nyquist frequency and p is the center-to-center distance of the pixels expressed in angular size for the eye. However, sampling does not only influence the integration limit, but also the part of the MTF below this limit. The MTF below this limit can be described by the following sinc function (Barten, 1991b):

$$M(u) = \left| \frac{\sin(\pi p u)}{\pi p u} \right| \qquad (9.10)$$

These expressions are valid for a monochrome display. For a color display, p has to be multiplied with a certain factor to replace p by the average sampling distance of the three colors.

The luminance deviations caused by the luminance quantization may be considered as a form of noise (Barten, 1993). The luminance distribution of this noise is in principle rectangular over the quantization distance ΔL between two luminance levels. As the measurements by Kayargadde given in the previous section have shown, a rectangular distribution of the noise is completely comparable with the usual Gaussian distribution, if the standard deviation of the distribution is used to characterize the noise. For the so obtained relative standard deviation σ_n of this noise can be derived (Barten, 1993)

$$\sigma_n = \frac{1}{\sqrt{12}} \frac{\Delta L}{\bar{L}} \qquad (9.11)$$

where $\bar{L}$ is the average luminance. The luminance levels used for the quantization are mostly equally spaced between zero and maximum luminance. For this situation

$$\Delta L = \frac{L_{max}}{N - 1} \qquad (9.12)$$

where L_{max} is the maximum luminance and N is the number of luminance levels. From the last two equations follows

$$\sigma_n = \frac{1}{\sqrt{12}} \frac{L_{max}/\bar{L}}{N - 1} \qquad (9.13)$$

In practice the ratio $L_{max}/\bar{L}$ in this expression can vary from about 1.25 for text images of black letters on a white background to 10 or more for natural images.

The spectral noise density can be calculated from σ_n with the aid of Eq. (2.41) given in Chapter 2:

$$\Phi_n = \frac{\sigma_n^2}{2u_{nmax} 2v_{nmax}} \qquad (9.14)$$

where u_{nmax} and v_{nmax} are the maximum spatial frequencies of the noise in x and y directions. These spatial frequencies are much higher than the maximum spatial frequencies u_{max} and v_{max} that can be derived from the pixel size p over which the luminance level is digitized. The truncation that takes place in the digitizing process creates higher harmonics of these frequencies. These frequencies are not visible in the displayed image but have to be taken into account in the calculation of the spectral noise density. From a mathematical analysis, we found that the following relations approximately hold for this process (Barten, 1993):

$$u_{nmax} = 3u_{max} \quad \text{and} \quad v_{nmax} = 3v_{max} \qquad (9.15)$$

where u_{nmax} and v_{nmax} are given by $1/(2p)$. From these equations follows for the spectral density of the noise caused by the luminance quantization

$$\Phi_n = \frac{\sigma_n^2 p^2}{3^2} \qquad (9.16)$$

With these equations the SQRI has been calculated for an experiment by Silverstein et al. (1990) where the subjective image quality was measured as a function of the number of quantization levels. The images were graphical images with different pixel geometries and different combinations of colors generated on a high resolution color CRT to simulate the effects that would appear on a matrix display. For this type of pictures a factor three may be assumed for the ratio between the maximum luminance and the average luminance. The viewing distance was 0.81 m and the field size was 1.5° x 1.5°. The simulated pixel density $1/p$ was 55 pixels/cm, so that p was 0.77 arc min. Ten subjects took part in the experiment. The measurement data were averaged over subjects, image types, colors, and two pixel geometries: triad and quad. To take the geometric effect of these color configurations into

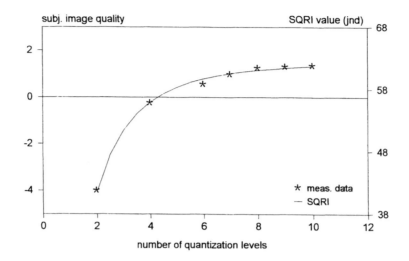

Figure 9.17: Subjective image quality as a function of the number of quantization levels measured by Silverstein et al. (1990) for graphical images simulated on a high resolution color CRT. Field size 1.5°×1.5°. The curve through the data has been calculated with the SQRI. The correlation between measurements and calculations is 99.7%. The image quality saturates at a number of about eight quantization levels.

account, the value of p in Eqs. (9.9) and (9.10) had to be multiplied with a factor 1.83 to obtain the average sampling distance for the three colors. Measurement data and calculation results are shown in Fig. 9.17. The correlation between measurements and calculations is 99.7%. From the figure, it appears that the dependence of image quality on the number of quantization levels is well described by the given model. It is further almost surprising to see that the image quality already saturates at about eight quantization levels (three bits).

In the experiment shown in Fig. 9.17, the pixel density $1/p$ was constant. In another experiment made by the same investigators (Kranz & Silverstein, 1990) the pixel density was varied besides the number of quantization levels. Twelve subjects took part in this experiment. Other conditions were the same as in the previous experiment. Measurements and calculations are shown in Fig. 9.18. The number of quantization levels is expressed here in bits. The measurement data are again averaged over subjects, image types, colors, and two pixel geometries: triad and quad. The correlation between measurements and calculations is 96.6%. The figure shows a good agreement between the measurements and the calculations.

In the design of an electronic imaging system, bits of luminance levels can often be exchanged with pixel density. For an optimum design, a good knowledge of the effects of both parameters on image quality is important. The analysis given here can be used to find an optimum trade-off between pixel density and bits of luminance

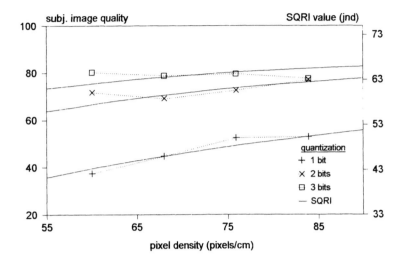

Figure 9.18: Subjective image quality as a function of pixel density measured by Kranz & Silverstein (1990) for graphical images simulated on a high resolution color CRT with three different numbers of quantization levels. The number of quantization levels is expressed in bits. Field size $1.5° \times 1.5°$. The solid curves have been calculated with the SQRI. The correlation between measurements and calculations is 96.6%.

levels. It must, however, be remarked that the occurrence of contouring effects in gradual luminance transitions, like skies, etc. is not taken into account in the model. For natural images, therefore, deviations from the calculated image quality may be expected. For such images, usually a much higher number of quantization levels is used than the three bits that were sufficient for the example given here. For these images, generally eight bits are used.

9.10 Summary and conclusions

In this chapter a survey has been given of the effect of various parameters on image quality. They cover areas of resolution, image size, viewing distance, luminance, contrast, gamma, noise, pixel density and luminance quantization. These effects have been analyzed with the aid of the SQRI metric given in the previous chapter, which appears to have a very good relation with subjectively perceived image quality. For the contrast sensitivity of the eye which is used in this metric, use was made of Eq. (3.26) given in Chapter 3. For special applications, like gamma, pixel density and luminance quantization, additional equations were introduced. In most cases, the calculated effect on image quality could be compared with published measurements. The results appeared to be in good agreement with the measurement data, also in situations where two different parameters were varied simultaneously.

References

Barten, P.G.J. (1987). The SQRI method: a new method for the evaluation of visible resolution on a display. *Proceedings of the SID*, **28**, 253-262.

Barten, P.G.J. (1989a). The square root integral (SQRI): a new metric to describe the effect of various display parameters on perceived image quality. *Human Vision, Visual Processing, and Digital Display I, Proc. SPIE*, **1077**, 73-82.

Barten, P.G.J. (1989b). Evaluation of CRT displays with the SQRI method. *Proceedings of the SID*, **30**, 9-14.

Barten, P.G.J. (1990a). Evaluation of subjective image quality with the square-root integral method. *Journal of the Optical Society of America A*, **7**, 2024-2031.

Barten, P.G.J. (1990b). Subjective image quality of high-definition television pictures. *Proceedings of the SID*, **31**, 239-243.

Barten, P.G.J. (1991a). Evaluation of the effect of noise on subjective image quality. *Human Vision, Visual Processing, and Digital Display II, Proc. SPIE*, **1453**, 2-15.

Barten, P.G.J. (1991b). Resolution of liquid-crystal displays. *SID Digest*, **22**, 772-775.

Barten, P.G.J. (1991c). The effect of glass transmission on the subjective image quality of CRT pictures. *Proceedings of the SID*, **32**, 285-288.

Barten, P.G.J. (1992). The SQRI as a measure for VDU image quality. *SID Digest*, **23**, 867-870.

Barten, P.G.J. (1993). Effects of quantization and pixel structure on the image quality of color matrix displays. *Journal of the SID*, **1**, 147-153.

Barten, P.G.J. (1996). Effect of gamma on subjective image quality. *SID Digest*, **27**, 421-424.

Feng, Y., Östberg, O., and Lindström B. (1990). MTFA as a measure for computer display image quality. *Displays*, **11**, 186-192.

Kayargadde, V. (1995). Feature extraction for image quality prediction. Ph.D. Thesis, Technical University of Eindhoven, Eindhoven, The Netherlands.

Kranz, J.H. & Silverstein, L.D. (1990). Color matrix display image quality: the effects of luminance and spatial sampling. *SID Digest*, **21**, 29-32.

Mitsubayashi, T., Shimodaira, Y., Washio, H., Ikeda, H., Muraoka, T., Mizushina, S. (1996). Gamma range of reproduced pictures having the quality above acceptable limits. *Proceedings Eurodisplay '96*, 483-486.

Nijenhuis, M.R.M. (1993). Sampling and interpolation of static images: a perceptual view. Ph.D. Thesis, Technical University of Eindhoven, Eindhoven, The Netherlands.

Östberg, O. & Feng, Y. (1991). The MTFA stick - an inexpensive measuring gauge for quick assessment of VDT display image resolution. *SID Digest*, **22**, 779-780.

Roufs, J.A.J., Koselka, V.J.F., and van Tongeren, A.A.A.M. (1994). Global brightness contrast and the effect on perceptual image quality. *Human Vision, Visual Processing, and Digital Display V, Proc. SPIE*, **2179**, 80-89.

Shimodaira, Y., Muraoka, T., Mizushina, S., Washio, H., Yamane, Y., Awane, K. (1995). Acceptable limits of gamma for a TFT-liquid crystal display on subjective evaluation of picture quality. *IEEE Transactions on Consumer Electronics*, **41**, 550-554.

Silverstein, L.D., Kranz, J.H., Gomer, F.E., Yeh, Y.Y., and Monty, R.W. (1990). Effects of spatial sampling and luminance quantization on the image quality of color matrix displays. *Journal of the Optical Society of America A*, **7**, 1955-1968.

Togerson, W.S. (1958). Theory and methods of scaling. John Wiley and Sons, New York.

van der Zee, E. & Boesten, M.H.W.A. (1980). The influence of luminance and size on the image quality of complex scenes. *IPO Annual Progress Report*, **15**, 69-75.

van Hateren, J.H. & van der Schaaf, A. (1996). Temporal properties of natural scenes. *Human Vision and Electronic Imaging, Proc. SPIE*, **2657**, 139-143.

Westerink, J.H.D.M. & Roufs, J.A.J. (1989). Subjective image quality as a function of viewing distance, resolution, and picture size. *SMPTE Journal*, **98**, 113-119.

Epilogue

In this book, a model has been given for the contrast sensitivity of the eye and its effects on perceived image quality. The model is based on the assumption that the contrast sensitivity is determined by internal noise in the visual system. In the different chapters, the model was extended to various aspects of the visual system. In all cases, the model predictions were compared with published measurement data. These measurements generally confirmed the validity of the assumptions that were made. At the end, the author would still like to make a few remarks that may be useful for further investigations.

Remarkable are some properties of the visual system that were met during the development of the model:
1. The constancy of the signal-to-noise ratio k and the reason of its rather high value of 3 (See section 2.2).
2. The low value of 3% for the quantum efficiency η (See section 3.4).
3. The limitation of the integration area of the eye by a maximum number of 15 cycles (See sections 2.4 and 4.4.2).

These properties should be investigated more deeply to get a better insight into the behavior of the visual system.

For practical reasons, the model was restricted to photopic vision (daylight vision) and also the vision of color was left out of consideration. Although the contrast sensitivity of the eye is mainly of importance at photopic viewing conditions and the perceived image quality is mainly determined by the achromatic properties of vision, an extension of the model to scotopic vision (night vision) and an investigation of the effects of color would be useful for further applications.

The constants used in the model are largely based on contrast sensitivity measurements with young adult observers between 18 and 28 years of age. The so obtained values are considered as typical values. It would be interesting to investigate how these constants change with age. Furthermore, it would be interesting to investigate the effect of visual defects, like cataract, glaucoma, macular degeneration, etc. on these constants. This could give a deeper insight into the visual system and the character of these defects.

For future measurements of the contrast sensitivity function, one should consider the following measures:

1. Using the psychometric function for determining the modulation threshold. This gives the most accurate measurement results.
2. Applying also external noise. This gives extra information of the constants that play a role in contrast sensitivity.
3. Using a constant field size and a constant viewing distance. By using a constant field size, other factors that influence the results, like pupil size and the used area of the retina, are constant.

For the given models a large number of assumptions had to be made about the biological structure of the retina and the visual processing in retinal elements and nerve fibers. It would be interesting to check these assumptions by direct biological and anatomical measurements. This would give a deeper insight into the visual system and could possibly support the validity of the models.

Subject index

Peter G. J. Barten graduated with a degree in physics from the Technical University of Delft. After military service, he worked at Philips in Eindhoven, where he was in charge of the development of color CRTs. After his retirement in 1987, he became an independent consultant, with special emphasis on image quality. He developed the SQRI method for the evaluation of perceived image quality and a model for the spatiotemporal contrast sensitivity of the human eye. He is the author of numerous technical papers and a chapter on Electron Optics in the *TV & Video Engineer's Reference Book*. He has taught seminars for the SID and short courses on MTF and image quality for SPIE. He is a fellow of the SID and a member of SPIE. In May 1999, he received his Ph.D. degree from the Technical University of Eindhoven for the subject of this book.